Bad Medicine

Misconceptions and Misuses Revealed, from Distance Healing to Vitamin O

CHRISTOPHER WANJEK

John Wiley & Sons, Inc.

For my father, Edward Leo Wanjek

CONTENTS

ACKNOWLEDGMENTS

Special thanks goes to Suzumi Yasutake, for her patience and her ample flow of tea; to Grace Mary Ott, Lee Carl, Steve Maran, and Skip Barker, who got it rolling, in that order; and, in refreshingly reversed alphabetical order, to the following for their help, advice, and/or ridicule: Michael Wanjek, Edward Wanjek Jr., Paul F. Tompkins, Richard Todaro, Craig Stoltz, Marc Stern, William Steigerwald, Jane Shure, Charles Seife, Eric Sabo, P. Barry Ryan, Richard Rinehart, Phil Plait, Charles Ott, Reba Novich, The National Institutes of Health, Amy Lanou, Ilana Harrus, Jeanne Gray, Thomas Graham, John Graham, Jeff Golick, Christopher Dickey, Amy Danzig, Beverly Cowart, Marie Coppola, Jessica Clark, Patrick Carroll, Carla Cantor, Howard Brightman III, Ann Bradley, Marin Allen; and to members of the Academy for voting for me.

Finally, special thanks to Dave Craigin and Jeff Lewis for the idea for Chapter 36.

INTRODUCTION

The Roots of Bad Medicine

Compared to the ancient cockroach, we truly are a baby species. The first humans rose above the tall grasses of central Africa a mere 150,000 years ago and, with their sharpened rocks and heightened curiosity, set forth to conquer the world. The cockroaches followed, never missing the opportunity for a free meal. Many millennia later, it's difficult to say who the winner is. Pound for pound, the world is weighted more with roaches than humans. And we humans are the hunted, easy prey for viruses and bacteria. This has not been an easy lesson for big-brained humans to learn—that we are not the dominant species; that there exists a world of microorganisms beyond what we can see with our eyes; that we are living not in the Age of Man but rather the Age of Bacteria. We didn't catch on for about 149,900 years, until roughly the close of the nineteenth century. But when we did . . . eureka! Germ theory. We immediately applied this hard-won knowledge to the field of medicine. We washed our hands, supplied clean water to cities, created vaccines, and understood the body in terms of cellular interaction. Suddenly, by the twentieth century, we were living, on average, at least twice as long as at any time in history.

All told, it is amazing we have come as far as we have. The forces of nature are well beyond our control and, at times, seem overwhelming. Drought and famine strike at will. Epidemics of disease wipe out entire cities and villages. Fires, floods, and earthquakes destroy in seconds what it took centuries to build. Imagine yourself forty thousand years ago, helpless. Let's face it, you're no

Einstein. Neither am I. I have no concept of how anything works. I have no knowledge of how to turn sand into glass, nor how to polish it in such a way as to make tiny objects appear larger. I have no impulse to do this. I have no reason to believe that anything exists smaller or more distant than what can be seen with my eyes. Sure, I'm curious as to why the sun makes rocks feel hot. The sun must be hot, I think. Still, I have no idea why the crops didn't grow this year, nor why eight of my ten children will die before reaching adulthood.

For countless millennia, humans as bright as you and I attributed both the good and bad in life to the gods. Life made more sense this way. Slowly, along the way, though, we learned to help ourselves. Certain plants—their fruits, roots, or flowers, maybe—seemed to make us feel better. Rubbing oils on the skin helped with horrible rashes and burns. Drinking teas helped soothe an upset stomach. Cedar ashes and incantations, however, didn't seem to help that massive head injury from a battle over water rights. But two working traditional cures out of three ain't bad. We didn't know why certain treatments worked and others didn't. We didn't ask why. We just did what we could to heal. Ultimately, we figured, the gods were calling the shots, allowing medicine to work.

Herbs have long been the medicine of choice. Wall drawings from around the world show the use of medicinal herbs as early as 40,000 years ago. The famed iceman Ötzi, whose near-perfect remains were found in the Italian Alps in 1991 after 5,000 years in a deep freeze, carried with him herbal cures for his stomach pains. His pharmacist, no doubt, was a spiritual healer. The spirits truly determined the fate of the healing process; thus the prescription of herbs and other cures was the domain of the priestess, shaman, or witch doctor. Healing relied on prayer and ceremony just as much as—if not more than—it did on the medicine itself. Life went on this way until about 3000 B.C.E. (before the common era).

The art of healing began to turn into a science in Egypt. China is often credited as having the oldest healing culture. This, of course, denies the existence of every other culture in the world. Five thousand years ago, China was no more ahead of the healing game than were the native cultures of the Americas and Australia.

The Egyptians, however, were starting to think about medicinal cause and effect by about 3000 B.C.E. More importantly, they were writing down their thoughts. The Egyptians determined that the heart was the center of thought, the liver produced blood, and the brain cooled the body. This is wrong, but it was a good start. Really, would you have any concept of what a brain or liver does without the battery of electronic probes at our disposal today?

Egyptian doctors applied herbs, surgery, and a little magic in a methodological approach to healing. Sekhet'eanach and Imhotep were among the first men who could be called doctors, as opposed to priests, herbalists, or witch doctors. They took note of their actions, determined what did and didn't work, and taught other doctors. One achievement from these early days was healing wounds with honey, which unbeknownst to the Egyptian doctors sealed the wound from outside infection and contained antiseptic agents as well. Proud like the rest of us, though, the Egyptian doctors assumed they weren't to blame when their best cures didn't work. The gods ultimately determined the fate of the patient.

In China, by about 2000 B.C.E., herbalists began recording which leaves, roots, and teas worked for various ailments. The Chinese didn't blame the gods so much for disease but rather imbalances in two sorts of energies, yin and yang. One could cure a disease by restoring that balance. This logic was a step beyond the Egyptians, who did not yet question what was causing the disease. The Chinese had several methods to restore balance and harmony. One was acupuncture and massage, which triggers the movement of a vital energy force called *qi* (chi) in order to push back the yin or the yang, whichever was causing the trouble. Exercises and breath work also got this *qi* moving. Herbs—containing the basic elements of fire, water, soil, wood, and metal—were considered the most potent means to affect yin and yang. These balancing methods were refined over the next two thousand years. Many of the herbs never worked; neither did they kill. So they survived to be written about in the medical books of modern times. The faulty logic of "like cures like" has perpetuated in this way, and many in China today still believe that ground tiger penis will cure human male impotency.

In India, by about 1000 B.C.E., doctors were performing surgery with success rates higher than anywhere else in the world at that time. Indian doctors knew how to drain fluids, sew wounds, remove kidney stones, and even perform simple plastic surgery (the punishment for adultery was having your nose cut off). Unlike China, though, healing practices were still intertwined with religion and ritual. India's Ayurveda system of herbs, diet, oils, and exercise developed a little later, around 200 B.C.E.

In the Western world, the Greeks picked up where the Egyptians left off. By around 400 B.C.E., Hippocrates was laying the foundation of modern medicine. Hippocrates was the first to subscribe to the notion that disease has a rational cause and therefore a rational cure. Out went the witch doctor and magic (for the time being, anyway). Hippocrates borrowed from China and India and established the concept of the four bodily humors: blood, phlegm, black bile, and yellow bile. He had a slight twist on the idea, though. Disease caused the imbalance in these fluids, which led to the symptoms of the disease. This was different from the notion that an imbalance caused disease. What caused the disease, Hippocrates reckoned, was poor diet, poor exercise, poor air, or that knife in your shoulder blade from the last battle.

By this point in history in the Western world, medicine was a full-fledged science. Hippocrates was right on the money in terms of diet, exercise, and fresh air. He regularly prescribed healthy foods and relaxation for his patients. His teachings inspired the Roman Empire to build intricate aqueducts, providing nearly every major Roman city in the Empire with fresh water, bathhouses, and sewage removal. Hundreds of thousands of people lived in cities by this point, and without such a system, disease would have been rampant. Hippocrates was dead wrong about the four humors, though; and he was a fan of bloodletting to eliminate an "overbalance" of blood. As we shall see later, medieval Europe decided to keep the humors and bloodletting but to ditch the part about clean air and food.

Claudius Galen was a Greek who was born in Turkey, trained in Egypt, and worked throughout the Mediterranean as the surgeon to the gladiators circa 150 C.E. He got around. Helping bruised and butchered gladiators, Galen knew the human body better than anyone else. Countless head injuries revealed that the brain was the center of thought, not the heart. Countless spinal cord injuries revealed that a nervous system controlled movement. Countless squirting veins revealed that the blood moves through the body. These were all major advances in our notion of how the body works. Galen built upon Hippocrates's theory of humors by introducing the law of opposites to treat disease. Fever, it was thought, was a result of too much hot yellow bile. Thus Galen prescribed cool liquids and cold food. Galen was also the first to map out anatomy. Working with cadavers in Rome was taboo, as it was in ancient Egypt, China, India, and Greece. So Galen studied pigs and created the first illustrated text of human parts, with the (false) assumption that pigs and humans were the same inside.

Surgery was still a daring and unperfected art. Doctors didn't understand the importance of sterilizing instruments and keeping wounds clean, and patients usually succumbed to infection. Amputations were sealed with hot iron, a deadly and painful procedure called cauterization. Cesarean births were performed only when the mother was dead or about to die. For this reason, we can determine that Julius Caesar was not born by cesarean birth, as legend and the *Oxford English Dictionary* have it. His mother was alive and well throughout Julius's reign, and no woman survived a cesarean birth until the eighteenth century.

Galen's teachings would dominate medicine for the next fifteen hundred years. The gist was this: Breathing brought a spirit called pneuma into the body from the ubiquitous "world spirit." Pneuma entered the body through the trachea and then through pulmonary veins to the heart, where it mixed with blood. Blood was not yet known to circulate, but it did slosh around with movement like water in a bottle. The pneuma, which we know today as oxygen, traveled through the body to produce activity. The brain, when infused with pneuma, initiated commands for movement. None of this is so far off, really. Rather good medicine, all in all. Of course, there was also the strange notion that the uterus was the cause of hysteria in women, and that the removal of the uterus (a hysterectomy) got rid of the hysteria.

Rome was sacked by the end of the fifth century, and rational thought moved west to the Arab world. There is little bad medicine to report from Persia and Arabia. The great doctors Rhazes (circa 900) and Avicenna (circa 1000) of Persia built upon the Greek tradition in applying scientific methods in cataloging diseases and treatments. Rhazes, or al-Razi, identified the difference between measles and smallpox. Avicenna, or Ibn Sina, was the first to notice that a dirty setting could infect a wound. The Holy Book of Islam, the Qu'ran, taught that the wealthy were responsible for the treatment of the sick and poor. By the twelfth century Baghdad was home to sixty hospitals, all free, compared to one hospital each in London and Paris at the time, both of which were beyond the means of the poor. And unlike European hospitals, Muslim hospitals from the Near East to Spain were inspected regularly and had separate wards for different diseases. Meanwhile, Asian medicine remained unchanged: Ayurveda in India, and herbs and acupuncture throughout East and Southeast Asia.

Back in the West, the medicine of Galen and Hippocrates mutated a bit over the centuries, and bad medicine emerged in Europe. This is the first time we can really use the term "bad medicine." Let's cut the ancients a break; they were just trying to figure things out. In the so-called dark ages, though, Europeans deliberately abandoned better therapies. The concept of sanitation and hygiene, established by the Greeks and Romans, was dismissed. So too the notion of rational causes for diseases. Here we have the root of bad medicine; and humans today—with their willingness to abandon vaccines, chlorinated water, and conventional medicine in favor of ancient cures—are entering into a personal dark age. Here's how it all got started.

THE FOUR HUMORS

The notion of the four bodily "humors" dominated Western thought from ancient Greece to nearly the twentieth century. Medicine in China and India was also based on a notion similar to the four humors. We just can't shake it. Most bad alternative medicine today, from Ayurveda and aromatherapy to touch therapy, is a direct throwback to the era of the humors.

The science of the four humors, although wrong, was never-theless brilliant in its thoroughness, and penetrated all aspects of life. The four humors, or bodily fluids, were blood, phlegm, yellow bile, and black bile. These corresponded to the basic elements of air, water, fire, and earth, respectively. Blood was hot, moist, and airlike. Phlegm was cold, moist, and waterlike. Yellow bile was hot, dry, and firelike. Black bile was cold, dry, and earthlike. These elements combined to make bodily fluids. Pus, for instance, was thought to be a combination of phlegm and yellow bile. Urine and feces were largely made of yellow and black bile, respectively. The humors also matched emotions and the four seasons. Blood was associated with spring (hot and wet) and being passionate or san-guine. Phlegm was associated with winter (cold and wet) and being phlegmatic, apathetic, pale, or downright cowardly. Yellow bile was associated with summer (hot and dry) and being choleric and violent. Black bile was associated with autumn (cold and dry) and being melancholy.

Foods, too, were characterized in this way in the medieval mind. Beef was hot and dry; black pepper was *very* hot and dry. Chicken, milk, and cheese were hot and wet. Root vegetables were cold and dry. Leafy vegetables and fish were cold and wet. Mush-rooms were *very* cold and wet, the complete opposite of black pep-per. Various degrees existed for the amount of dryness or coldness. This was the medieval equivalent of the four food groups, minus intense lobbying from the beef and dairy industries.

Humans, depending on their age and gender, were slightly hot and wet. Older folks tended to be a little colder and dryer. South-erners were more hot-blooded than northerners. Diseases resulted from an imbalance of the four humors, and the role of medicine was to get these humors in balance to cure that disease. One way was diet. A doctor would prescribe hot and dry food for a patient diagnosed as cold and wet. This meant no fruit or leafy greens, food that actually would have done the patient good. Likewise, a hot, wet person—whatever that might mean—got cold, dry root vegetables.

The other way to restore balance was through those notorious purges. Bloodletting was the way to rid the body of hot, wet humor. Doctors must have assumed that most of their patients

were too hot and wet, for bloodletting was by far the most common treatment for most diseases for about two thousand years in
Europe. At best, bloodletting could lower a fever for these patients.
But the procedure, as you can imagine, had little therapeutic effect.
(Withdrawing a small amount of blood is useful in the treatment
of polycythemia, an excess of red blood cells, although doctors did
not know of this disease until the twentieth century.) Blood, after
all, is kind of important. A body is most vulnerable to disease
when the blood level is low, for less blood means fewer disease-
fighting white blood cells. Even inflammation, which may look like
a swelling of too much blood, is aggravated by blood loss. Yet
most wealthy Europeans subjected themselves to a bimonthly
bloodletting as a form of preventive medicine. Bleeding was important to them because they thought the liver makes blood from
food, and thus the body needs to get rid of excess blood from
excess food consumption. Doctors would drain several pints of
blood, often stopping only when the patient passed out. Barbers,
with their access to razors, were key players in the bloodletting
business—how much medical training does one need to slice open
an arm? The barbershop was marked on the outside with a red
and white poll, as it is today, representing white bandages around
a bloody, red arm.

Bloodletting was a way of life in the Western world. The break-
through in bloodletting technology came with the advent of leech
therapy. Leeches allowed for controlled bleeding, and doctors of
great learning could prescribe specific numbers of leeches for specific body parts for a specific duration. In France in the early 1800s,
hospitals regularly subjected patients to leeches before doctors even
saw them. Inspired by Dr. François Broussais, who proposed that
all disease resulted in an excess buildup of blood, the city of Paris
went through six million leeches annually by the 1830s, leading to
the commercial extinction of leeches in France. Leech therapy was
certainly less painful than traditional bloodletting but equally as
useless and harmful.

Bloodletting, or phlebotomy, as the discerning medieval scholar
was apt to say, was but one of many purging therapies. Remember, the idea was to restore balance, and a good way to do that

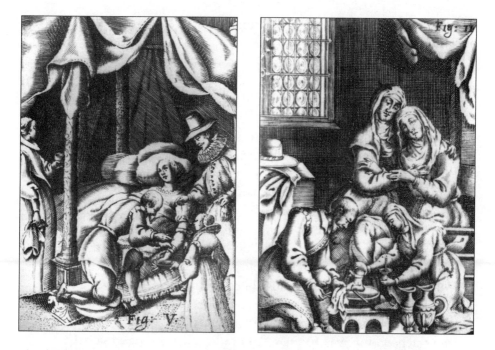

Blood was taken from the arms, legs, hands, feet, tongue, or rump. This went on for about two thousand years. Illustrations by Cintio d'Amato. *Courtesy of the National Library of Medicine*

was to get rid of the excess humor that was causing your blues. Step 1 was to determine the excess humor, be it blood, phlegm, yellow bile, or black bile. Diagnosis was open to broad interpretation based in part on the patient's mood, color of skin or tongue, smell or taste of urine, and reported diet, as well as time of year and astrological events. Alternative healers today often use these techniques as well. Purges promoted bleeding, vomiting, sweating, blistering, expectoration, and defecation. These were usually accomplished by some type of toxin. The poisonous lobelia herb, called the vomit weed, had obvious results. Poultices of dung induced sweating and encouraged coughing. A poisonous white powder called tartar emetic was valued as an expectorant as well as for its vomit-inducing properties. Mercury compounds served as laxatives and suppositories. And a litany of nasty plants caused blisters to form on the skin. Please be assured that this was all natural medicine.

Purging continued right up to the end of the nineteenth century in Europe and in the United States. Benjamin Rush, a signer of the American Declaration of Independence, was a pioneer in bloodletting—and also concentrated his efforts on discouraging masturbation by making it painful for American soldiers (see chapter 3). Rush successfully sued a prominent journalist in Philadelphia for suggesting that he was a quack who killed most of his patients with excessive bloodletting. Apparently the courts favored purges, too. Legend has it that George Washington died from a bloodletting. He actually died of acute bacterial epiglottitis, an inflammation of the little flap that blocks food from entering the lungs when swallowing. The 80 ounces of blood withdrawn over 12 hours to relieve that inflammation most likely didn't help matters. (One of Washington's doctors suggested a tracheotomy to ease Washington's breathing, but he was overruled in favor of bloodletting, for the tracheotomy was too new and radical a form of surgery.)

The industrial age merely brought horrible new methods to elicit humors from the body. Industrial acids became the blistering agents of choice. Mercury was a common additive in medicines because of its unique, seemingly magical liquid form at room temperature. Mercury is fourteen times heavier than water, so it was used either to flush out the digestive system or, as the theory of the day stated, to open up blocked channels in the body to allow for better flow and mixing of humors. Electricity, a new toy, was tested on all body parts—yes, all of them—to get those humors flowing and to cause a blister or two. Rev. John Wesley, whose life spanned most of the eighteenth century, was a pioneer in electrotherapy. To his credit, he hoped for a new treatment that would eliminate the need for painful purges and complicated, poisonous concoctions. Undergoing mild electrical shocks to unblock "humor channels" was probably less painful than other purges, but alas, just as useless.

It is one of the great mysteries of humankind that bloodletting and other purges could continue for so long with no clear benefit and only adverse affects. Simple trial-and-error demonstrates that purges do harm. Equally strange was homeopathy's rise in popularity through the nineteenth century. Homeopathy is essentially the prescribing of sugar or salt water, although the original practitioners didn't know this at the time. There's no medicine in homeopathy, but the cures did no harm. Regardless of the disease, sugar water will do less harm than bloodletting and blistering every time.

Today, remnants of the four humors are seen in Ayurveda, aromatherapy, and other nonsensical, noneffective alternative therapies. Ayurveda, from India, is based on diets and herbs that fit personality types, stellar alignment, and the basic elements of fire, water, and air. Aromatherapy incorporates the concepts of mood, imbalance, the zodiac signs, and the alignment of stars in choosing the right combination of essential oils to burn. Magnet therapy and crystal healing borrow elements of the "cold and dry" rock idea to heal excess "hot and moist" humors in the body. Touch therapy uses the so-called positive energy in one's hands to unblock so-called energy channels in the body and engender the flow of humors and restore balance. These therapies are collectively known as holistic, as all ancient therapies were, because they treat disease not as a single bacterium or virus but rather as an imbalance in the greater workings of the body as a whole.

So if bloodletting seems terribly outdated, you can always experiment with forms of modern-day holistic medicine, which many health insurance companies now cover. Holistic medicine is often less expensive than conventional treatment, and, in the long run, it is cheaper for health insurance companies because truly ill patients die more quickly. Sad but true.

MEDIEVAL SUPERSTITION

In alternative medicine the phrase "practiced for centuries" really means "steeped in superstition." Medieval medicine was an odd mix of superstition and science that survived the fall of the Roman Empire by word of mouth. Ancient books survived largely in the Arab world and in select Christian monasteries. Most of Europe had little knowledge of, let alone access to, the writings of Hippocrates, Galen, Pliny, and the other founding fathers of scientific medical thought. The notion that disease had rational causes and rational cures went out the window. Disease, once again, was caused by God's wrath, human sin, or evil spells.

Just as the Egyptians, long before Hippocrates, worshipped the legendary, godlike healers Imhotep and Asclepius, Europeans turned to Christian saints to cure them. The list of saints is quite extensive. The beautiful Saint Lucy, having plucked out her eyes to render herself unattractive to lustful male passersby, became the patron saint of eye diseases. Saint Blaise, having saved a child from choking on a fishbone, became the patron saint of throat diseases and coughs. And so on. Often, the only medical treatment available was prayer, which, again, worked better than purges.

Highly touted today as alternatives to conventional medicine, herbal remedies can work. In fact, some can work "as is," such as black cohosh root for menopausal hot flashes. Some need to be refined: A chemical in the bark of the willow tree is the active ingredient in aspirin; a chemical in the rauwolfia herb, when isolated and highly concentrated, works as a tranquilizer. Yet many herbs have no effective medicinal properties. These herbs were taken hand in hand with hemlock and mercury through the ages merely because they apparently worked once. No one challenged the authority of the priest or doctor administering the herb. Garlic did not ward off the plague. And despite what modern herbal remedy books say, calamus, toxic in small doses, does not cure deafness or epilepsy. Comfrey, toxic in small doses, does not cure ulcers. Nutmeg, toxic as a whole seed, is not a sedative or an aphrodisiac. The list of legendary, useless herbs is sizeable. These herbal remedies are not time-tested, as the modern books claim. Rather, the herbs were associated with some local or national hero or saint, prescribed in certain seasons, during certain stellar or planetary alignments, for patients of certain zodiac signs who were exhibiting a certain characteristic of one of the four bodily humors.

Tapping into ancient cures is tantamount to tapping into the faulty logic behind the cures and the notion of disease. Most people assumed that the black plague was caused by witches, Jews, or some sinful person. Likewise, dirty looks caused birth defects and lots of other diseases. A pregnant woman who gazed at a butchered

TABLE I—The Four Humors

The Four Humors of Ancient and Medieval History		The Four Humors of the 21st Century	
Phlegm— winter, apathetic	Blood— spring, sanguine	Sarcasm	Irony
Yellow bile— summer, choleric	Black bile— autumn, melancholy	Whoopie cushion	Aromatherapy Ayurveda

animal could give birth to a child with a cleft lip. Cause and effect: I gazed, and look what happened. Laws were enacted from Scandinavia to southern Germany prohibiting butchers from hanging rabbits and certain other animals in their storefronts. This is the scientific foundation of alternative medicine.

THE AGE OF REASON

But before we get too cocky, we must remember that we still know so little. The Age of Reason and Enlightenment in the eighteenth century was, in many ways, a return to the dark ages once again by virtue of narrow-mindedness. Gems of this period include the invention of race, with Europeans being the smartest; eugenics, the notion that forced sterilization could prevent alcoholism, criminal behavior, and general stupidity; phrenology, the study of head shapes to determine intelligence and personality; antisex and anti-masturbation crusades to improve health by controlling impulses and conserving energy and fluids; and charlatan medicine men and true quacks selling cure-all tonics created with so-called modern chemical techniques.

The most potent pill born of this era, still ingrained in Western thought, is the notion that heavy-duty science can and will conquer all. Thus, we create medications such as statins to control high cholesterol levels when exercise and diet work to prevent high cholesterol in the first place. We search for the gene causing obesity or heart disease or whatever when lifestyle factors, not genetics, are the main determinants of these diseases. We use a barrage of antibacterial agents to clean or treat mild infections when the body can do the job just as well. With all of these therapies, however effective, come side effects. The all-out pursuit of scientific perfection sends many citizens of the industrialized world scurrying for alternative cures, which leads them into the arms of bad medicine.

So here we are at the dawn of the twenty-first century. We have come a long way, and we have a long way to go. Bloodletting is gone; but a common method of treating cancer is chemotherapy, which weakens the whole body along with the cancer. The all-but-certain death from surgery is gone; but we still have tens of thousands of deaths each year in the hospital due to infections and medical error. Malnutrition and vitamin deficiencies are largely behind us in the industrialized world; but obesity is killing us, and poorer countries are still malnourished. The human genome reveals we are all one people; but racism still exists and markedly affects health. Technologies exist to keep drinking water safe, food plentiful, and viruses at bay; yet we choose to abandon them and undermine the great public health achievements of the twentieth century—just as medieval Europe abandoned the teachings of the Ancients. Today, will we continue to march forward or will we take steps back? Our fate, perhaps, depends on our ability to recognize bad medicine.

PART I

I Sing the Body Eclectic

Your absence of mind we have borne, till your presence of body came to be called in question by it.

—Charles Lamb (1775–1834)

Our concept of the human body comes from myriad sources—anecdotes, legends, and old wives' tales, some old and some new. We have a better understanding of the little things, such as how a virus infiltrates the defense system of the body. We all dutifully load up on medicine and vitamins to fight a battle we cannot see but only fantasize about. Yet when it comes to the large scale—tastes on our tongues or issues of race and brain size—we overlook the obvious. We miss the forest for the trees, or perhaps the body for the cells.

1

❦ ❦ ❦ ❦ ❦ ❦

10 Percent Misconception, 90 Percent Misdirection: The Brain at Work

Often it is said that we use only 10 percent of our brain. Is the brain really a vast, untapped resource of incomprehensible powers? Absolutely. I've heard countless vapid cell-phone conversations on street corners that attest to this. I remember one young lady giddy over a "brown baby pigeon" that was hopping about her feet while she was talking to her friend. The bird was a sparrow.

Remarkably, she was using nearly 100 percent of her brain in describing the "baby pigeon." Optic nerves were relaying the image of a tiny brown bird to the visual cortex way in the back of her brain via the thalamus, sort of the brain's relay station. Cochlear nerves in her ears were transmitting the electrical impulses of the sound of her friend's inane chatter through the brain stem and thalamus to the auditory cortex, where it was ultimately interpreted as language in her brain's Wernicke's area. Memory is spread widely through the brain, from the hippocampus and amygdala to the cerebral cortex, so it is not clear where the young lady was accessing the incorrect information that small brown birds in the city are baby pigeons and not sparrows. Most certainly, though, her brain stem was relaying motor function from her cerebellum and cerebral cortex to the muscles, enabling her to hold the cell phone, turn her head, unconsciously check out a cute guy, and more or less to stand and breathe. Her brain's hypothalamus was regulating her body temperature. All and all, it was a busy time for her brain.

Our budding ornithologist might not have been using a full 100 percent of her brain on the cell phone all at once. After all, no one exercise utilizes 100 percent of one's muscle system. But she was using far more than 10 percent. More importantly, by the time she woke up in the morning after dreams of baby pigeons and cute guys, she would have used all of her brain. All of the brain's regions and many of its neurons would have gotten a workout.

Now, how you use your brain is your own business. You can read *War and Peace* or you can watch dating shows on television. While many argue that the latter is a waste of the brain's potential, no one can justifiably say that 90 percent of the brain lies dormant, like some untapped oil well, waiting to gush forth with unrealized brilliance.

The "10 percent" brain myth goes back at least a hundred years, perhaps more if one considers the teachings of transcendental meditation and the concept of maximizing the mind's power. Albert Einstein, whom no one accused of having a lazy brain, may have helped keep the myth alive when he told a reporter, wryly and perhaps sarcastically, that his brilliance came from using more than 10 percent of his brain. But this tale cannot be confirmed. Barry Beyerstein, a neurologist at Simon Fraser University in British Columbia, tried to isolate the origin of this myth in "Whence Cometh the Myth that We Only Use Ten Percent of Our Brain?," a chapter in the book *Mind Myths: Exploring Popular Assumptions About the Mind and Brain*. Beyerstein finds reference to a "silent cortex" in brain studies from the 1930s, as well as seeds of misconception from the 1800s.

The nineteenth century was a time of remarkable advancements in our understanding of the physical and biological world. The French physiologist Pierre Flourens's groundbreaking work on the brains of rabbits and pigeons in the 1820s and 1830s mapped out regions in the brain responsible for basic movements, memory, and mood. Basically, he removed parts of their brains and took notes on what the animals could no longer do. A few decades later, Pierre Paul Broca, a French physician, isolated the region in the human brain responsible for controlling speech. He performed autopsies on stroke victims who had lost the ability to form words (but could still comprehend language). In the 1870s, Gustav Fritsch and Eduard Hitzig, two German physiologists, improved upon Flourens's work by zapping certain regions in a dog's brain with electricity and seeing which muscles moved.

The electrical zapping continued with greater precision in the 1930s. Researchers found that in all their brain volunteers, from animals to humans, there were certain regions in the brain that did not respond to stimuli. These regions were labeled the "silent cortex," and humans had a lot of them. The name was not meant to imply that the regions were inactive; merely, the electrical stimuli didn't provoke anything obvious, such as twitching. Further research has shown that the "silent cortex" is responsible for the very traits that make us human: language and abstract thought.

How can we be certain that we don't use only 10 percent of the brain? As Beyerstein succinctly says, "The armamentarium of modern neuroscience decisively repudiates this notion." CAT, PET and MRI scans, along with a battery of other tests, show that there are no inactive regions of the brain, even during sleep. Neuroscientists regularly hook up patients to these devices and ask them to do math problems, listen to music, paint, or do whatever they please. Certain regions of the brain fire up with activity depending on what task is performed. The scans catch all this activity; the entire brain has been mapped in this way.

Further debunking the myth is the fact that the brain, like any other body part, must be used to remain healthy. If your leg remains in a cast for a month, it wilts. A 90-percent brain inactivity rate would result in 90 percent of the brain rapidly deteriorating. Unused neurons (brain cells) would shrivel and die. Clearly, this doesn't happen in healthy individuals. In Alzheimer's disease, there is a diffuse 10 percent to 20 percent loss of neurons. This has a devastating effect on memory and consciousness. A person would be comatose if 90 percent of the brain—any 90 percent—were inactive.

The "10 percent" brain myth is silly even from an evolutionary standpoint. The brain is a hungry organ, requiring energy (in the form of oxygen and glucose) all day and all night. This organ, comprising only 5 percent of the body's total weight, consumes 20 percent of the oxygen and glucose. Evolution would have never favored a big, useless "high-maintenance" brain if only 10 percent of it were vital for survival. Darwin aside, just use common sense. Never do we hear a doctor say, "Fortunately the bullet wound destroyed the 90 percent of the brain he doesn't use. He's good to go; call me in the morning."

True, there are bizarre brain stories: people impaled by lead pipes and, still functioning, suddenly taking up an interest in yodeling; or people who have up to half their brain removed to control seizures. The brain never truly recovers its full capacity in these situations, but it can learn to adapt—particularly if the patient is young. The brain can reroute its wiring, or neural pathways, to maintain most of its function. Children whose parts of their brain have been damaged or removed can grow up, if treated, to lead productive and seemingly normal lives. Adults with brain damage have far greater difficulty attaining full function. This is because their streets have already been paved, unlike a child who is growing and learning. It is easier to pave a new street around a damaged area than it is to rip up an old street and start anew.

Yoga masters—and often those who are paralyzed from the neck down—learn how to better control their autonomic nervous system, that part of the nervous system responsible for things we do automatically without "thinking," such as breathing and regulating blood flow. For example, you are walking down a dark street and suddenly a mugger jumps in front of you with a knife. Your heart starts pounding. The rise in heart rate is a result of the sympathetic autonomic nervous system, the fight-or-flight response. Conversely, the parasympathetic autonomic nervous system will lower your heart rate and metabolism rate, allowing your body to conserve energy during times of rest. When you control your autonomic nervous system with your brain, you are not using any new brain parts. You are simply more conscious about using sections of the brain you have used all your life. Yoga masters have been known to lower their pulse rate well into the 30s, compared to a resting pulse rate of 70 or so for most other people. Paralyzed individuals can learn how to regulate their bowels, and, in the case of men, even achieve penile erection by controlling autonomic nerves with their brain. But none of this is the unused 90 percent that psychics and other frauds talk about.

The "10 percent" figure popped up somewhere in the twentieth century. At first, the language was nonspecific, with lines such as "Scientists say we don't use most of our brain's power." In 1944 an ad for the Pelman Institute, which offered self-improvement courses, appearing on the inside front cover of a wartime Penguin edition of Stella Gibbons's novel *Cold Comfort Farm,* was perhaps one of the first to nail down a number:

> What's holding you back? Just one fact—one scientific fact. That is all. Because, as Science says, you are using only one-tenth of your real brain-power!

This is where the psychics and believers in extrasensory perception (ESP) pick up the ball. The mantra of those people who harness the Force as adeptly as Luke Skywalker is that your "other 90 percent" of the brain has the power to sense and move what the mundane 10 percent cannot. Uri "Sorry, I can't bend this spoon in a controlled laboratory setting" Geller is a magician who claims to use his brain to move objects without touching them and to read other people's minds. He's quite successful. With his clever brain, Geller mysteriously convinces fools to reach into their wallets and fork over big bucks to buy his books and to watch him perform. He's a consummate mind reader, knowing what his audience will fall for. In the introduction to his 1996 book, *Mind Power,* he writes:

> [M]ost of us only use about 10 per cent of our brains, if that. . . .
> I believe that we once had full power over our minds. We had to, in order to survive, but as our world has become more sophisticated and complex we have forgotten many of the abilities we once had.

Makes sense to me: the proliferation of books, quantum mechanics, superconductivity, semiconductors, laser surgery, X-ray telescopes that can probe black-hole event horizons . . . all these things are making us stupid! Me hunt, me eat. That's the kind of stimuli we need. I will build shelter and a fire with my ability to mind-bend this spoon. Why is it that Geller can use his mind power to bend a spoon and not a lever in a Coke machine to get a free drink? Beats me. I must be part of the 10-percent-and-under crowd.

One cannot even speak of 10 percent in a diffuse sense, that our brains are only 10 percent full of knowledge. There's no limit to the mind's ability to store knowledge. This would be like saying we use only 10 percent of our ears because we never listen to 90 percent of the world languages, or 10 percent of our taste buds because we never eat 90 percent of the foods that others eat.

Metaphorically, this great brain tithing is a reflection of our deep-seated human inferiority complex: ancient civilizations could not have accomplished what they did on their own, we say; there must have been aliens guiding them or they must have moved massive stones with their minds. If Einstein could determine that mass distorts space in such a way to produce gravity, we say, he must have had access to a different part of the brain than I do. However, we cannot ignore the core message of the Uri Gellers and the fraudulent psychics—that humans often fail to attain their fullest potential. We can, as a species, rise above the ignorance of bigotry or fraud or malice, not by tapping into unused mystic portions of our brains but by reveling in the pursuit of knowledge.

Well, maybe tomorrow. There's a rerun of *Married with Children* on the tube.

2

𝍏 𝍏 𝍏 𝍏 𝍏 𝍏 𝍏

Big Brain, Little Smarts:
Brain Size and Intelligence

In Kurt Vonnegut's novel *Galapagos,* big-brained humans blow up the world with nuclear weapons. The only survivors are cruise-ship passengers shipwrecked on one of the Galápagos Islands of Darwin fame. Survival of the fittest plays out on the island, with those able to catch fish better suited to eat, live, mate, and pass on their genetic information. Smart people—the kind who can build weapons that destroy the world—are at a disadvantage on the island because all they know how to do is argue. They soon die. The dumb people, over the course of millions of years, evolve into dumber, penguinlike creatures skilled at catching fish. Vonnegut clearly doesn't have much respect for those with big brains. By "big brain," of course, he means the so-called smart person—creative liberty from a great author who knows deep down that human brain size has nothing to do with intelligence.

Assuming you could measure smartness (which we can't), and assuming you could measure brain size by measuring the outside of the head (which we can't), you'd still be wrong to assume that people with bigger heads are smarter. There have been geniuses with tiny brains and idiots with huge ones. Women have smaller brains than men, on average. Smaller people, particularly midgets, often have smaller brains. Unless you are prepared to defend the stance that women and short people are dumber, you'd be wise to drop the "big brain = big smarts" argument.

23

If the brain were a muscle, you'd be right in assuming that a
bigger brain means more mental strength. Yet the brain is far more
complicated than a muscle. The brain is a fluid-rich, spongelike tissue containing ten billion nerve endings controlling every thought
and movement we undertake. The notion that a big brain equals
big intelligence goes back several hundred years, yet it was in
ancient times that humans first began to identify the brain as the
organ that controls thought. The concept wasn't so straightforward. Imagine yourself with no medical instruments. How can you
tell that the brain—which you see when you slaughter an animal—
is responsible for thought in humans? Aristotle, a noted smart guy,
thought the brain was a radiator that cooled the blood. The center
of thought was the heart, according to Aristotle. This was around
350 B.C.E. Around 150 C.E., Galen, famed doctor to the Roman
gladiators, began to noticed that violent head injuries from ridicu-
lously gory gladiator games led to neurological disorders. He suggested that the brain might harbor thought, a concept met with
giggles.

Barbarians of all brain sizes sacked Rome late in the fifth century, and serious thought went underground for a while. The philosopher René Descartes revisited the brain in the seventeenth century. Descartes, of "I think, therefore I am" fame, suggested that
mental activity took place in the soul and transmitted itself to the
brain, which served as a transceiver of thought. He was quite ada-
mant that the brain was just a relayer and not the location of mental activity. A few hundred years later, phrenology suddenly
became the rage. Phrenology is the study of head shapes to determine intelligence and personality. Phrenologists from Europe were
the first group to subscribe to the idea that smart people have big
brains and that other races were dumber because of their supposedly smaller heads.

Mind you, no group of people have smaller heads than others.
In his book *The Mismeasures of Man,* the Harvard geologist and
noted evolutionist Stephen Jay Gould reviewed data from centuries
past to show that head measurements across races are more or less
the same. Often, inaccuracies in measurements were a result of
either foolishness or fraud, two fixtures of bad medicine that are

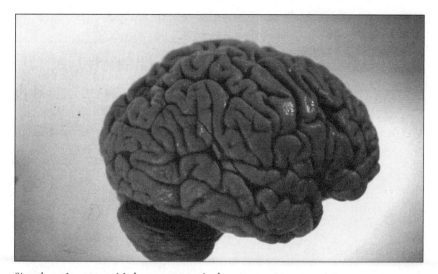

Size doesn't matter; it's how you use it that counts. *Courtesy of the National Institute of Neurological Disorders and Stroke*

difficult to discriminate. In one experiment from the nineteenth century, two skulls—one from an Englishman and the other from an African—were filled with gravel. The Victorian scientists packed gravel into the English skull and loosely filled the African skull, apparently demonstrating that English skulls hold more gravel and, therefore, larger brains. At any rate, the experiment said a thing or two about who had rocks in their heads.

Today, white supremacist groups and eugenicists—those who seek selective mating to produce superior offspring—use poor Gould's chart of brain sizes to show that they really do vary by race. (And again, even if they did—which they don't—this has nothing to do with intelligence.) The charts in Gould's book show that northern Asians have the largest brains, followed closely by Europeans. Native Americans and southern Asians have smaller brains. Ancient Europeans had even smaller brains, and modern Africans have the smallest. The problem here is the sampling. The size differences are small: 87 cubic inches for modern Europeans versus 83 cubic inches for modern Africans—although eugenicists argue this is the difference of millions of precious brain neurons. That may be true, but other samples of brain sizes show Africans having larger brains than Europeans. It all depends on your sample population, and early headhunters collected the heads that best supported their arguments of Caucasian superiority. Phrenology was in full swing. Americans and Europeans alike used this pseudoscience as justification for the slave trade and the killing of native peoples in the Americas and Australia.

But what of big brains? Women have smaller brains compared to men. Are they dumber? Easy now. The average brain size is about 3 pounds or 1,400 grams. The brain of the French writer Anatole France was only 2.24 pounds, well below average. Lord Byron's brain was nearly twice this amount, over four pounds. These two geniuses with vastly different brain sizes lived roughly in the same era. Albert Einstein had an average-sized brain, most likely the same size as yours and mine. You can't even compare humans to other animals. Dolphins have about the same size brain as humans. Elephants' brains are five times bigger. Whale brains are bigger yet. If you compare the ratio of brain mass to body mass, the rat is the winner. Maybe rats *are* smarter. You try navigating the New York subway in the dark.

This all comes down to what is unique about the human brain. Whales and elephants need huge brains not to think but to move. Most of the whale's enormous brain, up to ten times bigger than a human brain, is devoted to moving its massive fins and sensing feeling along its massive body.

The human brain is unique in that it has a highly developed section called the cerebral cortex, which is located in the frontal lobe of the brain. The cerebral cortex is essential for processing thought and language. Early humanoids had a less developed cerebral cortex and therefore could not attain what we commonly call conscious experience. The same can be said for modern apes and dolphins. An ape's brain could get bigger, but unless the cerebral cortex develops in a certain way, the ape will never achieve "thought." The cerebral cortex is merely one section of the brain. A dog's brain has a larger section devoted to smell, and therefore dogs can detect and remember smells better than humans, regardless of brain size. Dogs went one way, humans went another.

Scientists are far from understanding what constitutes the "mind"—that combination of skills responsible for decision making, emotion, perception, imagination, and self-awareness. Conscious experience does not arise from one neuron, nor is it confined to the cerebral cortex. The "mind" seems to be a neural network, a hardwiring of brain nerve cells with each cell connected to fifty thousand of its neighbors. Smarter people—creative, scientific, or physically skilled—make better use of the human brain through networking. Size doesn't matter, but how one relays nerve impulses around the brain does. Drug addicts and alcoholics hinder their ability to think by damaging neural networks. A connection is broken, and a skill or memory is lost. Likewise, neurological disorders such as Alzheimer's disease involve broken networks.

The brain of a child is primed for hardwiring, yet humans can generate neural connections throughout life. Taxi drivers in London, for example, develop over the course of many years a larger hippocampus, the section of the brain responsible for navigating and remembering directions. This well-circulated finding, conducted on about two dozen taxi drivers, sure gave cabbies a big head. The study confirmed the theory that certain types of thought lead to greater development of a particular part of the brain. With this development come more neurons, more capillaries, more blood, and, yes, more mass—in the case of the taxi drivers, a good milligram or two in a 1,400-gram (1,400,000 mg) brain. Inactivity in other parts of the brain leads to shrinkage. Overall, though, the brain doesn't gain much mass by "thinking hard."

Some of us are born with a brain better designed for certain types of thought. The brain is like farmland. True geniuses—which are few and far between—are often those people with one section of the brain that is more fertile than others. Einstein, for example, had a larger inferior parietal region, the part of the brain responsible for mathematical thought and the ability to visualize movement in space. This section was 15 percent wider, perhaps at the cost of making another section smaller (possibly the hair-combing section). Also, Einstein's brain lacked a groove called the sulcus that normally runs through this part of this brain. This absence may have allowed the neurons on either side to communicate more easily.

The bottom line is that Einstein's brain was just different, not larger. If eugenicists had their way, they would not "breed" other Einsteins, because Einstein had an average-sized brain. By selectively choosing big brains and big brains only, you would miss the brain of an Einstein, of an Anatole France, and of the countless great artists, musicians, thinkers, comedians, and hard-working ordinary folks with average-sized brains or smaller.

Evolutionists have no problem accepting the fact that brain size doesn't matter. We say that humans became humans as their brains got bigger, but this is only partially true. Yes, early humanoids had smaller brains. As the prehuman developed, it grew larger but, more importantly, progressed in such a way that allowed for thought—separating humans from every other animal. Whale brains got bigger and bigger, too, as whales got bigger. Whales didn't necessarily become smarter because of it.

The human brain, by the way, isn't getting any bigger, nor are humans getting innately smarter. We are no smarter than the cavemen, those clever souls who domesticated fire and figured out that grain makes bread. Admit it. It took ingenuity to melt certain rocks into copper, bronze, and iron. A caveman alive today, socialized as a modern human, would be just as smart or dumb as the rest of us, depending on your perspective.

Humans will get smarter in terms of learning new things, despite the potential for permanent stupidity from watching television. Humans will build upon the knowledge of preceding generations. We will understand new physics and create technologies beyond our comprehension today. We may very well master deep-space travel and discover new dimensions and forces in the universe. Our brains will stay the same size, though. The notion of a future human with an enormous head to house an enormous brain is pure fantasy. Evolution simply doesn't favor larger heads over small heads. Evolution doesn't even favor smart people over dumb people. Dumb people mate with stellar success. For humans to develop bigger heads, we would have to kill off people with small heads and only mate with large-headed people. Of the offspring, only the largest of the large heads could mate. Then, over tens of thousands of years, assuming this ridiculous practice of big-head mating continued, humans would have larger heads. What we would gain is uncertain. Baseball caps would need to stretch; this much is sure.

3

℞ ℞ ℞ ℞ ℞ ℞

Blinded by Lies: The Eyes Have It

Myths about vision loss are easy to believe because they seem rather logical. Children who sit close to the television are often fitted for glasses. Reading in poor lighting strains your eyes, making words blurry. Adults who spend hours in front of a computer screen eventually find themselves making their first visit to an eye doctor. However, few everyday activities in this modern world will lead to vision loss. This time, your mother was wrong.

What we are witnessing is a slight confusion between cause and effect. There just may be a reason why a child is sitting so close to the television. That child might be nearsighted, unable to see at a distance. It's only natural to move closer. Likewise, a nearsighted child may hold a book very close to her face. The act of sitting or reading too close did not cause the nearsightedness. They were the result of a vision problem.

Activities bring about problems we never knew we had. Dyslexia, for example, didn't show its face until humans invented written language. Similarly, those who do not read will never come to the realization that they need glasses to view tiny letters. A bookworm wears eyeglasses because he needs them to read. The reading didn't cause damage to his eyes. Rather, the necessity of reading caused the realization that this bookworm would need a particular form of magnification in order to see the tiny letters printed on a page. Reading is one of the most focused tasks your eyes will accomplish. Until a few hundred years ago, most people were illiterate. So what is the gauge for "poor" eyesight? Up to about age forty, most people have vision decent enough to perform daily nonreading tasks such as farming and chopping. But consider that bookworm living a couple hundred years ago. He starts reading at ten, and he doesn't need glasses. By age forty he might need them due to a general decline in vision caused by aging.

Had that bookworm stayed on the farm and planted turnips, he would have lived a life devoid of eyeglasses. Perhaps his vision would have diminished over the years, but never to the point where he could no longer see turnips. Tiny print, yes. Yet what concern is tiny print to an illiterate farmer? Centuries ago, only book-smart people wore eyeglasses, because they were the only ones who needed them to read fine print. Logically, the misconception spread that years of scholarly life (in dim, flickering candlelit dormitories) led to vision problems—because scholars wore glasses and farmers didn't.

Onward to the twenty-first century. The hours we spend in front of a computer are not causing eye problems. Rather, the eye problems we have were there all along, lying dormant until we started working on a computer. You may be able to read a newspaper without reading glasses. You keep the paper no farther than twelve inches from your eyes when you read it, and you usually read a newspaper in decent lighting, perhaps under a lamp in a comfortable chair. Yet most folks keep about twenty-four inches between a computer screen and their eyes. The print is tiny, the letters flicker, and the monitor itself emits a glare that makes the words hard to read. After months of using a computer, folks who thought they had good vision (able to read a newspaper without eyeglasses) find that they cannot read a computer screen. The first assumption is that the computer screen caused the vision problem, when in reality from the very beginning these people had the type of eyes that couldn't read tiny, flickering print twenty-four inches away. Few people do. Computers are hard to read. This is why electronic books have limited appeal.

Any difficulty you encounter reading a newspaper at night after working on a computer all day is the result of muscle strain, not eye damage. Eight hours on a computer is hard work for your eyes. Rest for a few days, and you'll see that you can read a newspaper just as you always have. The same is true for reading under poor lighting. Reading in the dark is like biking uphill: both cause muscle strain. You can bike longer on flat terrain and read longer in proper lighting. When lights are dim, your eye muscles strain to let in more light. Reading becomes more difficult and words become fuzzy. Rest up and you will be fine the next day, with no damage to your vision from the night before.

The same logic applies to wearing eyeglasses: they don't make your eyes weak or dependent on glasses. If you can no longer read at all without your eyeglasses, your vision has gotten worse on its own. Eyeglasses improve your vision, but they don't "heal" your eyes or stop the general decline in vision associated with old age.

Can we do anything to keep our vision strong? Aside from not sticking or spraying things in your eyes, the best thing you can do is have your eyes examined for cataracts, glaucoma, and other eye ailments, which are usually treatable when detected early. Most of us will need eyeglasses by retirement age for at least some activities, such as reading. The lens of the eye and the muscles controlling it weaken over time, and there is no eye exercise that can prevent this. A gradual decline in vision is natural. Diabetes is unnatural. Diabetes is a disease that can lead to poor vision. So, staying healthy and maintaining a normal body weight, which minimizes your risk of developing diabetes, ultimately protects your eyes. In America, diabetes poses by far the greatest risk to vision.

There is evidence, tenuous at best, that reading tiny print for long hours will strain your muscles and lead to permanent eye damage. Most eye doctors say that you would be too tired to read any further long before you destroyed your muscles from excessive reading. So-called "near work" is another issue. Workers laboring in sweatshops—stitching clothes or soldering computer chips—often do develop eye problems because the nonstop work is straining their eye muscles beyond repair. The eyes are controlled by muscles, and even a young baseball pitcher will throw out his shoulder and develop permanent nerve and muscle damage if he throws a ball too hard and too long. Yet aside from this, you can make your eyes hurt, but you can't hurt your eyes.

You cannot buy eye protection in a bottle. It is true that a diet deficient in certain nutrients, particularly vitamin A, can lead to blindness. You need to be pretty low on A, though. This is no longer a problem in the developed world. The converse—that consuming more "eye" nutrients will give you better vision—seems to be mere wishful thinking. Studies have yet to show a connection between diet and super vision. The largest study to date, funded by the National Eye Institute and reported in October 2001, found that people with intermediate-stage, age-related macular degeneration (AMD) could reduce their risk of progressing into an advanced stage and going blind with a high-dose combination of antioxidants and zinc. This certainly is good news for those with AMD, a disease of the retina and a leading cause of blindness among the elderly. Yet those same nutrients did nothing to prevent AMD, slow its progression in early stages, improve vision, or prevent cataracts. How about carrots, the good-vision vegetables of lore? Carrots are rich in beta-carotene, which the body uses to manufacture the antioxidant vitamin A. Any healthy diet will give the body enough vitamin A, with or without the help of carrots. In other words, carrots won't improve your vision.

The latest vision-nutrient craze is lutein, added to multivitamins or sold by itself. Lutein is said to slow down or even prevent and reverse age-related vision loss, as well as stave off cataracts. As with all of the assertions you read on vitamin bottles, the lutein claim has not been scientifically validated. At best, it's just a hunch. The lutein craze didn't materialize out of thin air, though, like some other far crazier claims mentioned in this book. Lutein is a yellowish pigment found naturally in the eyes and in leafy green vegetables, such as spinach and kale. If lutein is already in the eyes protecting them, then ingesting more of it would certainly be a good thing, one would think. Well, the dietary supplement industry thinks that. Let's investigate.

Lutein is an antioxidant. The first part of the lutein theory is that its antioxidant properties prevent chemicals in the body called free radicals from damaging the cells that make up the retina. You'll read in Part IV, however, that the whole antioxidant theory is on shaky ground. Forget about that angle for now. The second part of the lutein theory is that lutein acts as a natural eyeshade, protecting the retina from ultraviolet radiation, the type of light that causes sunburn. Lutein is one of three pigments covering the macula, a region of heightened sensitivity in the retina. The disease AMD is associated with a breakdown of pigment. So getting more pigment (i.e., more lutein) might make the situation better, right? That's the million-dollar question, and the answer seems to be no. Studies have shown that only about half the people who take lutein supplements actually end up with more lutein in the eyes. And even then, the extra lutein doesn't seem to help vision. If lutein were so great, you would think that doctors would add lutein directly to the retina. Yet they rely on complicated surgical procedures for treating AMD, which should lead you to believe that lutein supplements won't work.

Bilberry is another popular dietary supplement for good vision. Legend has it that Royal Air Force pilots in World War II spread bilberry jam on their morning toast, and as a result had better vision to bomb targets in German cities at night. The Allied Forces won the war, so the stuff must have worked. Bilberry has never been validated scientifically to improve vision. The British pilots were indeed on target, a result of improved radar devices developed for the war—which remained clean of any bilberry jam.

Perhaps no vision myth is as farfetched as the notion that one will go blind from masturbation. If this were true, we'd all be blind. Surely sexual intercourse, which is crucial for a species' survival, won't cause blindness. How can the eyes "feel" the difference between foreplay, coitus, and going solo?

The masturbation myth grew out of real fears that male masturbation was tantamount to the loss of precious seed needed to produce more humans. The Judeo-Christian tradition has always given masturbation a bad rap, even though it is not mentioned specifically in the Bible. (In Genesis, Onan spilled his seed while having sex with his brother's widow, not masturbating, and the Lord promptly killed him.) Science in the eighteenth century, amazingly, supported this notion that masturbation was physically harmful, even for women, although many scientists (all men) were not convinced that women masturbated with enough regularity to be of concern. The Swiss scientist Simon Andre Tissot was one of the first to declare, at least "scientifically," that masturbation leads to blindness in the 1758 *Treatise on the Diseases Produced by Onanism*. The idea was that the general waste of bodily energy led to a weakened state. This was an expansion of ideas raised in the anonymous sensation *Onania; or the Heinous Sin of Self-Pollution*, from 1717. Specifically, Tissot claimed that surging blood-flows from any type of sexual activity caused pressures on the body, but masturbation was particularly bad, causing fiercer surges that ultimately weakened the fragile blood vessels in the eyes.

Western nations took Tissot seriously. The newly formed United States of America couldn't afford to have a nation of blind citizenry. Benjamin Rush, a physician and a signer of the Declaration of Independence, was particularly influenced by Tissot and railed against masturbation in the fledgling states. Antimasturbation devices became readily available for those wishing to cure themselves of the disease of self-pollution. Devices included tubes with metal spikes that fit over the penis, for bedtime use, or other contraptions that made erections painful.

During the nineteenth century, Sylvester Graham and John Harvey Kellogg tried to squelch sexual desires through diet with new recipes for crackers and cereal, respectively. Kellogg also suggested circumcision without anesthesia for the chronic male masturbator. The two men, moral and health crusaders, were very successful in convincing the government and the public at large that masturbation was at the root of poor physical and mental health. (Along with blindness, Kellogg attributed acne and sleepiness to masturbation; just what every teenage boy wants to hear.) Sadly, the masturbation myth continued well into the twentieth century; the practice was referred to as a "functional and nervous disorder" in most medical textbooks until the 1950s. That's when studies from Masters and Johnson found that the vast majority of American adults masturbated and the vast majority of American adults, as far as they could see, could see.

4

ꙮ ꙮ ꙮ ꙮ ꙮ ꙮ ꙮ

All in Good Taste:
How the Tongue Works

Tongues are wagging in the science community about how the tongue senses taste. Oddly enough, more is known about vision and hearing, much more complicated senses. The last twenty-five years, however, has brought forth a wealth of tasteful understanding. Most researchers now agree that there are at least *five* basic tastes, not four. Taste buds are located on the tongue, the back roof of the mouth, and the throat. Remember that "tongue map" with the sweet taste buds on the tongue's tip and the salty taste buds on the side? That's all wrong.

Western society has traditionally described taste in terms of four qualities: salty, sweet, sour, and bitter. This jibed well with the concept of the four bodily humors. All other tastes would be a combination of these basic tastes, the theory went. Do these basic tastes really work like the primary colors? Researchers aren't sure. The human eye has three types of photoreceptors that work in unison to translate the broad range of wavelengths of visible light into the colorful hues of the rainbow. Taste buds, though, seem to work on their own. So it is not clear whether taste is an analytic sense, with a different receptor for each taste, or a synthetic sense like vision, with combinations of basic tastes producing unique new tastes. If the former is true, then there must be more types of receptors than just the "basic four" to account for the rainbow of tastes.

The Japanese have a word for a distinct fifth taste, called *umami*, which is the taste of glutamate. *Kombu*, a type of sea vegetable similar to kelp, has this *umami* taste; and the Japanese commonly use *kombu* in soup stocks or as a side dish. Scientists have indeed found an *umami* taste bud, which detects glutamate and other amino acids. Americans might recognize the taste of *umami* in monosodium glutamate, or MSG, a flavor enhancer. Researchers also may soon add fat to the list of tastes. Fat was long thought to be a sensation of smell and touch, not taste, with the creaminess of the fat brightening certain pleasure chambers in the brain. Yet fat substitutes, equally creamy, do not taste as nice as the real thing. From an evolutionary standpoint, early humans would have benefited from a fat taste receptor, for fat is a good source of long-term energy and warmth and serves as storage and transport for vitamins A, D, E, and K. Interestingly, Aristotle suggested that fat might be a basic taste; but then again, he recommended goat urine to cure baldness.

While the true number of basic tastes is debated, it is clear to researchers that the ubiquitous tongue map—which graces many a doctor's waiting room and is widespread today in science textbooks—is based on a hundred-year-old misinterpretation. You know the map: the taste buds for "sweet" are on the front of the tongue; the "salt" taste buds are on either side of the front of the tongue; the "sour" taste buds are in the middle of the tongue; and the "bitter" taste buds are way in the back. The map has frustrated many a grade-schooler who couldn't get the taste-test experiment right in science class. They couldn't get it right because the teachers were wrong. (I myself failed for insisting I could taste sugar in the back of my tongue.)

The tongue map is easy enough to prove wrong at home. Place salt on the tip of your tongue. You'll taste salt. For reasons unknown, scientists never challenged this for over a hundred years. It all started in the nineteenth century with a Dr. D. P. Hanig of Germany, who set out to measure the relative sensitivity on the tongue for the four basic tastes. (There was no Japanese market in town for *kombu* broth.) Hanig had lots of volunteers, and he placed drops of various sweet, sour, bitter, and salty liquids at different locations on their tongues. He then plotted his results. On average, the maximum spot for sweetness was on the tip of the tongue and the minimum was in the back. Bitterness maxed out on the back of the tongue. Saltiness had equal sensitivity across the tongue. Hanig concluded that sensitivity to the four tastes varied around the tongue. That's all.

In 1942, Edwin Boring, a noted psychology historian at Harvard University, took Hanig's raw data and calculated real numbers for the levels of sensitivity. Even these numbers merely denoted relative sensitivities. For sweetness, the tip of the tongue registered at 1 on Boring's scale; the back of the tongue had a sensitivity rating of 0.3. So the tip of the tongue is about three times more sensitive at detecting sweetness than the back of the tongue. That's all. This doesn't say that the back of the tongue cannot be a decent sensor of sweets.

Many scientists misinterpreted Hanig's and Boring's work and assumed areas of lower sensitivity were areas of no sensitivity. The tongue-map myth was born. In 1974, Dr. Virginia Collings reexamined Hanig's work and agreed with his main point: There were variations in sensitivity to four basic tastes around the tongue. But the variations were small. The four tastes can be detected any place on the tongue where there are taste receptors. And these receptors are everywhere—around the tongue, on the back roof of the mouth (called the soft palate), and even in the throat (which is strange because by this point you've already made the decision to eat this food that you may not be liking). Then there are the quasi–taste buds. Nerves in your tongue sense the soft texture of fat and send tastelike messages to your brain about what you are eating. Receptors on your eyeballs sense the "taste" of chili peppers. Your nose takes in the odor of food, which, when combined with the sense of taste, creates the concept of flavor.

Much of what we think we are tasting with our tongues we are really sensing with our noses. Hold your nose and eat a piece of chocolate. Chances are you won't taste "chocolate." You might taste "sweet" and "bitter," but without the sense of smell (and often vision), you won't know what that bittersweet sensation is. The flavor of Lifesavers candy is also difficult to guess without smell and the visual clues of food coloring. If you ever had a fierce head cold with a stuffed-up nose, this should come as no surprise. Food is tasteless when you have a cold because you cannot smell; your taste buds are fine. The complete absence of the sense of taste—known as ageusia, analogous to blindness and deafness—is far more uncommon than anosmia, an inability to smell. There is no good estimate on the prevalence of ageusia and anosmia in the United States, but about 5 percent of the population suffer from some degree of chemosensory dysfunction, as these conditions are called. The cause is usually nasal or sinus disease, allergies, viral damage, or head trauma. The Monell Chemical Sense Center in Philadelphia says that of the 1,200 or so patients evaluated over a 15-year period, only 5 (0.4%) were truly ageusic and another 5 suffered from substantial, generalized taste loss. Almost a third of all the patients had profound or complete smell loss.

Chances are, if you cannot taste, it is because you are not smelling properly. Now, if you're still thinking to buy that purple velvet painting of Elvis, then you have a whole other type of taste problem.

As for the myth that the tongue is the strongest muscle in the body, this doesn't seem to be true by any popular definition of strength. When considering pure strength, the greatest externally measured force produced by one muscle, then the masseter wins. This is the jaw muscle, and there are two, one on either side of the jaw. The masseter has an advantage over other muscles in that it is broadly attached to the jawbones, which act as levers. So the firm attachment plus a mechanical advantage make the jaw muscles the strongest. The world-record clench, according to the Guinness book, is 975 pounds for two seconds. No single quadricep muscle could produce that much force; a combination of muscles is needed. Now, if you consider an equal attachment to bone but with no mechanical advantage, then the quadriceps and the good ol' gluteus maximus win the strength game. These striated muscles have the highest concentration of fibers, a pure measure of strength. The heart is the strongest muscle if you measure "strength" as continuous activity without fatigue. The tongue wears out quickly . . . at least with some people.

5

ঽ ঽ ঽ ঽ ঽ ঽ

Scrubbing Your Liver:
The Demystification
of Detoxification

Scores of dietary supplements and herbal health remedies claim they can detoxify the liver. By "detoxify," they mean scrub it clean, give it the old Lysol treatment, wash away all the pollutants of the modern world. Sounds logical. After all, the liver filters the blood, finding harmful chemicals and breaking them down into less harmful chemicals. But this process does not result in an organ that is saturated with toxins needing to be banged clean like a lint screen. What the liver cannot detoxify it passes on through. All these products that call themselves liver detoxifiers have nothing to detoxify. The liver ain't toxic.

The potential for confusion lies in the fact that the liver itself is the detoxifier, not the herbal remedies. Everything you swallow that is broken down and absorbed into the bloodstream passes through the liver. The body depends on the liver to regulate, synthesize, and secrete many important proteins and nutrients and to purify, transform, and clear toxic or unneeded substances. Detoxification is the process of turning potentially harmful chemicals (from alcohol, medicines, or even food) into water-soluble chemicals that, by the end of the treatment, are usually less toxic than the original compound and can then be safely excreted. The idea is to break down as many toxins as possible. The fewer harmful chemicals circulating

39

through the body, the better. That's a no-brainer for the brain, which commands the liver. The liver cannot detoxify everything; toxins do slip through. Ultimately, these toxins are excreted from the body, with or without the once-over from the liver. The liver simply ensures there are fewer toxins available to wreak havoc in the body at any given time.

Those toxins that do get by the liver unscathed or merely wounded may pass through again. Or, if they are water soluble (as opposed to fat soluble), the kidney can break them down. Regardless, they do not stay in the liver. Vitamin A, when ingested in very large amounts, can accumulate in the liver and cause problems; and iron and copper can accumulate there due to rare genetic conditions. But that's about it.

Of course, none of this should suggest that dietary supplements and herbals are not helpful to the liver. In fact, these remedies may help the liver cope, bolstering it during times of trouble. This could come in handy for some. Years of alcohol abuse will leave the liver in a sorry state, unable to detoxify even lightweight toxins. Certain types of cholesterol-lowering drugs called statins, when taken by some people, can leave the liver a little weak, too. Studies are not yet conclusive, but some herbals show great promise not in *detoxifying* the liver, as they claim, but rather in helping the liver do the detoxifying.

The herb milk thistle, for one, is well accepted in Germany as an effective drug in treating death-cup mushroom poisoning. (Not that the Germans are always right; homeopathy and urinology are also popular there.) The death cup, as the ominous name implies, does a nasty number on the liver and soon moves on to the central nervous system. Milk thistle's active ingredient, silymarin, is nearly 100 percent effective in treating people who eat this common poisonous mushroom. In America, where silymarin has not been approved in medicinal form, the survival rate for death-cup poisoning is below 30 percent. Studies in Germany indicate that silymarin works by quickly protecting and restoring liver cell integrity. Studies are now under way in Europe on the effectiveness of medicinal silymarin and milk thistle itself in treating alcohol-induced liver damage. This herbal wouldn't be detoxifying the liver. Rather, it would heal the liver so that the organ can do the job it was intended to do.

According to the National Liver Foundation, there are no specific foods or herbals known to be healthy to the liver. Their recommendation? Same as everyone else's: eat plenty of fruits and vegetables, drink lots of water, and get plenty of physical exercise. The National Institutes of Health is just beginning to fund research on herbals and liver disease. One study, to be completed by the end of 2002, is investigating Vietnamese and Chinese herbals for the treatment of cirrhosis in rats. (No figures are available concerning the incidence of alcoholism in rats.) Foods purported to be healthy for the liver, although never proven, include dandelion, beets, and Japanese shijimi clams. Dandelion, the nemesis of suburban lawn enthusiasts, is actually one of the healthiest greens around, loaded with potassium, calcium, and vitamin C.

Be careful of the dietary supplements, though. Some supplements contain a "proprietary blend" of ingredients, protected by trade law. Consumers and physicians alike cannot know the percentages of the ingredients. They are not tested or approved by the FDA because they are "natural"—as natural, perhaps, as orange juice or poison ivy. Nature goes both ways. Many liver supplements contain niacin, which may be hepatotoxic, or damaging to the liver, in high doses. Many supplements also contain vitamin B_{12}, which is often found in excess amounts in patients with liver disease. If you have liver problems, you should check with a doctor. If you don't have liver problems, you should be protecting your liver through diet and exercise, not a potentially harmful and expensive, completely irrelevant, liver detox scheme. We're talking up to $20 for a ten-day supply. Although the liver has a remarkable ability to regenerate itself, your pocketbook may not.

6

ፄ ፄ ፄ ፄ ፄ ፄ

Refer to the Appendix:
Useless Organ or Helpful Player?

Enlightened scientists have said lots of dumb things over the past two hundred years. Most physicists in the 1880s, for example, lamented the fact that their field had learned all there was to learn about the laws of nature. They didn't even know about X rays at this point, let alone quantum mechanics. Earlier in the 1900s, biologists surmised that the human body had over one hundred useless parts left over from our more apelike lifestyle of a few million years ago. The parathyroid was one such organ, now known to regulate calcium-phosphorous metabolism. The appendix was another. Smart doctors today know better than to say they know everything.

Countless biology textbooks still say the appendix is useless, but nothing could be further from the truth. The appendix is a slimy, dead-end sac that hangs between the small and large intestines. It's about a half inch in diameter and three inches long. This organ certainly had a more prestigious role many, many years ago, when humans weren't human. Some primates, in fact, have hardworking appendices. Scientists think the appendix helps primates today, and helped prehumans way back when, to digest fiber and raw meat. Hard-to-digest food goes into the sac, and "good" bacteria along with bodily secretions start to break it down.

Our bodies changed with the course of evolution, but you can't keep a good organ down. The tiny, developing appendix in a tiny, developing fetus starts making endocrine cells around eleven weeks after conception. Endocrine cells secrete useful chemicals, such as hormones. The endocrine cells in the appendix secrete juicy chemicals (namely, amines and peptide hormones) that help with biological checks and balances as the fetus grows. After birth, the appendix mainly helps the body stave off disease by serving as a lymphoid organ. These organs, with their lymphoid tissue, make white blood cells and antibodies. The modern human appendix, by virtue of its lymphoid tissue, has become part of a complicated chain that makes B lymphocytes (one variety of white blood cell) and a class of antibodies known as immunoglobulin A antibodies. The appendix also produces certain chemicals that help direct the white blood cells to the parts of the body where they are needed most.

The dirty old gut is a good training ground for young white blood cells that will soon go forth and kill foreign invaders. The appendix, routinely collecting and expelling foodstuffs in and out of the intestines, exposes the white blood cells to myriad bacteria, viruses, drugs, and bad food present in the gastrointestinal tract. In this way, the white blood cells can learn to fight potentially deadly bacteria, such as E. coli. The appendix's contribution to the body's white blood cell and antibody production reaches its peak when you are about twenty or thirty years old; then production falls off sharply.

By age sixty, your appendix serves very little active purpose, but, passively, it serves as a nice spare part during surgery. Yes, the appendix can cause trouble when food gets stuck in there. This food essentially rots, causing an infection. The infection can be fatal, particularly if the appendix bursts. Once infected, the appendix needs to be removed. Life goes on; you'll likely never miss it. In the not-too-distant past, zealous doctors would remove the appendix during other types of surgery—to get it "out of the way" just in case it would some day become infected. The surgeon would say: the appendix is useless; I'm already elbow-deep into this person's gut; why don't I just snip the appendix now? But no more. Doctors now realize they can use the appendix for reconstructive surgery. In one type of bladder replacement surgery, for example, doctors take part of the intestine to form a bladder and use the appendix tissue to re-create a sphincter muscle, which can contract and open the bladder during urination. Similarly, the appendix is used as a substitute ureter, a tube that carries urine from the kidneys to the bladder.

Admittedly, the appendix isn't the most important chicken in the coop, but don't sell it short. You can also get by with one kidney or eye, after all. The more we learn about the body, the more we understand that everything has a purpose . . . even, believe in or not, the brain, although it's hard to tell with some people.

7

Going Gray? Not Today: White Hair and Its Causes

No doubt hundreds of ghost stories have a character whose hair has turned white from fear. Even the titans of literature have passed along this myth as if it were medical fact. So does Lord Byron in the 1816 poem *The Prisoner of Chillon:* "My hair is gray, but not with years, / Nor grew it white / In a single night, / As men's have grown from sudden fears." The concept is bold and terrifying enough. Imagine a fright so chilling that you age countless decades in a single night. But with apologies to Byron, it just ain't so.

As widespread as the myth is, there has never been a documented case of a person whose hair has suddenly turned white from fear or any stimulus other than hair dye. Certainly there are legends. Thomas More's hair was said to have turned completely white the night before his execution in 1535. His tumultuous final days as advisor to King Henry VIII are well documented in the annals of English history and dramatized in Robert Bolt's play *A Man for All Seasons.* The bit about hair turning white, however, came after he died—sort of like George Washington chopping down the cherry tree. The same type of legend surrounds the beheading of Marie Antoinette. Her hair was indeed white when she died, but most likely it had turned white slowly, months or years before the execution. (Another legend states that Marie

Antoinette's hair turned white on her daring yet failed escape from France.)

Most people go gray slowly, over the course of decades, as hair follicle after hair follicle starts producing gray hair. We have 100,000 or so hair follicles, so this can take some time. Usually, once a follicle starts producing gray hair, it doesn't turn back. Year by year more and more follicles make the switch until the entire head is gray. People have gone gray in a few months, not from fear but from the normal process of aging. By mechanisms unknown, all the hair follicles decide to start producing gray hair at roughly the same time. In a few months, when the colored tips are cut away only the gray is left.

You cannot go white or gray any faster than in a month unless you keep your hair extremely short. The root of each hair resides in the follicle, which is like a hair factory. If that hair is lost, a new one will grow in the same follicle. Cells in the follicle produce the protein keratin, the main ingredient in hair. Follicle cells called melanocytes make melanin, the same pigment that gives skin its color. Melanin colors the keratin. Folks who are flush with melanin have dark hair; those with little melanin have blond hair. As we age, melanocytes stop producing melanin. Without this pigment, the hair becomes white or grayish, which, by the way, is often an optical illusion produced by white hair mixing with darker hairs. This is a gradual process. The whiteness starts from the root and only moves as fast as the hair grows. Nothing sends "whiteness" along the hair strand. Hair is a lifeless strand of keratin that cannot transport nutrients or information up and down the strand. Hair growth is essentially the addition of new keratin to the root end of the strand of hair. When a black-haired person "goes gray," the new keratin added to the root end of a black strand is white. A day goes by, and the hair grows longer with the addition of more white keratin at the root end of the black strand. Soon a barber cuts the tips of the hair, the part that is black, and all that is left is the white underneath.

If you have very short hair, perhaps a military cut, you can go gray or white quickly because your half-inch of youthful hair (black, brown, blond, or red) is cut away, leaving a week's growth of new, white hair. Some people have white hair that goes unnoticed. The addition of a few new strands of white hair, however, may tip the scale, giving the impression that such a person suddenly went white. Hair fully grows out in about seven months. So it would take at least that long, on average, for all the colored hair to fall out and all new white hair to come in.

There is a rare form of sudden baldness, called diffuse alopecia areata. With this malady, only pigmented hair falls out, leaving gray and white hairs behind. The inattentive casual observer might assume that a person afflicted with diffuse alopecia areata has suddenly gone completely white. Look closer and you'll see that person has also lost half of his or her hair. This can happen as quickly as in a couple of weeks, and the hair loss can be stress related. But we're talking about losing clumps of hair, not a smooth and sudden transition from lush color to lush whiteness.

How did the myth get started? It could be that time flies faster than you think. You may think a friend has gone gray suddenly, when in reality you haven't seen him in a year. Marie Antoinette was under arrest for quite a while and was out of the public eye. On the day of her public execution, if Marie Antoinette's hair was white, one could easily assume it had just turned white. The townspeople knew her only with dark hair. She goes into prison with dark hair, comes out with white hair. The fear of death must have done it. I myself only "knew" folk singer Arlo Guthrie from pictures on his early albums from the 1960s and 1970s, such as *Alice's Restaurant* and *Amigo*. When I saw him perform live in 1991 and saw that mane of long, white hair, it seemed as though Arlo had aged overnight. Wonderfully enough, after his first song, Arlo even said: "I know what all of you are thinking: 'Man, he got old.' Well, guess what? You got old, too."

8

ʘ ʘ ʘ ʘ ʘ ʘ ʘ

Samson's Delight: Baldness Cures

Feminists are amused by the quest to find a cure for baldness, as if it were a disease. Millions of dollars are spent each year on baldness research, and this, many women say, is a prime example of how men control the flow of medical research funding. Can you blame men, though? Over 50 percent are bald or have significantly thinning hair by age 50. Year by year it gets worse: 30 percent are balding at age 30, 40 percent are balding at age 40, and so on. Women often mock men for being vain about their lost locks, but, truth be told, 20 percent of women have thinning hair and 5 percent lose it in clumps around the crown, just like men . . . and it's a real big deal for them, too.

Very soon, perhaps within a decade, there will be a drug that spurs head hair growth. At first it will most likely have some ridiculous side effect, like impotence or raised blood pressure. Then, after a few more years, the kinks in the drug will be worked out and then all we'll have to worry about are kinks in our hair. Researchers know what causes hair to stop growing, and pharmaceutical companies are pouring millions of dollars into drug development because they know the antibaldness pill will be as big as Viagra.

Of the thousands upon thousands of baldness cures to hit the market, only two have been approved. Minoxidil was introduced as Rogaine in 1988. This topical solution keeps thinning hair from falling out through, believe it or not, some unknown process. Minoxidil was first a blood-pressure medication; hair growth (in unwanted places) was a side effect. Finasteride entered the arena in the 1990s, marketed as a pill called Propecia or, in higher doses, Proscar. Taken orally, these pills retain hair that would have been lost by inhibiting an enzyme responsible for balding. These drugs have to be taken continuously, or hair loss returns. Before 1988, there was nothing but snake oil, figuratively and literally. Baldness cures have a long, smelly, messy history. Thanks to the Internet, they're all coming back.

Baldness has always had a bad rap. Countless references in the Bible note that God will make Israel's enemies either bald and sterile, bald and confused, bald and feeble, or just plain bald. In Revelation, at the end of the world, God will render select groups of evil people bald. You all know about Samson, whose great strength was in his hair. The (bald) prophet Elisha no doubt took that story too personally, as relayed in the Second Book of Kings. On the road out of Jericho, a group of boys made fun of Elisha, shouting, "Get out of here, baldy." Elisha cursed them so that "two she-bears came out of the woods and tore forty-two of the boys to pieces." His ways are mysterious.

The ancient Egyptians were among the first to develop treatments for baldness—rancid fat from snakes, geese, crocodiles, hippos, lions, and/or ibexes. These were serious topical ointments, no greasy kids' stuff. The bad smell was key, for it was proof that the concoctions were working. We're still fooled by this today, for everyone knows medicine is supposed to taste bad. And remember that dandruff shampoo: it tingles!

The great doctor, Hippocrates, treated his patients' baldness with pigeon droppings and other crap. Aristotle, as brilliant as he was, tried goat urine to address his own baldness. Julius Caesar was bald, which is ironic, because the name Caesar, from the Latin *caesaries,* means "abundant hair." His gal, Cleopatra, prepared pastes for him made of ground horse teeth and deer marrow. (Rancid hippo fat was clearly passé in the Egyptian courts by this time.) Alas, Cleopatra's salves didn't work. Neither did Roman cures of sulfur, tar, and the finest samples of animal urine from around the Mediterranean. Julius apparently caved in and simply tried to cover his bare head; he took to wearing wreaths of laurel. (He was known as the king of bad combovers, too, according to the Roman scribe

Pliny.) The mighty Hannibal was bald and didn't like it one bit. Like Captain Kirk in *Star Trek,* Hannibal was never in a fight scene without his toupee.

Baldness treatments such as urine and rancid fat survived the fall of the Roman Empire, unlike those worthless, pagan tomes on geometry and iambic pentameter. The Renaissance brought cow saliva. (Ah, cow *saliva,* not cow urine. Progress.) Meanwhile, in China, treatments moved forward with the introduction of animal testes mixed with ground herbs. Meditation and headstands had long been a standard cure there and in India. Finally, with the advent of modern technology in the late 1800s, baldness treatment entered the realm of the titillating: electric shock, vibrators, motorized scalp massagers, and suction devices.

What do all these treatments have in common, aside from the potential of making you look silly? They all work on three premises: increasing blood flow to the scalp, unclogging pores or hair follicles, and providing nutrients. Maybe these treatments really do all that, but they still aren't treating the causes of baldness, though. Baldness, for the most part, is purely genetic. You'd have to be literally starving to lose your hair due to poor nutrition. This is certainly possible, but far from likely. You don't need extra blood up there, either. The head is already flush with blood. The brain kind of needs blood, and the body makes sure to deliver it through two relatively massive arteries in your neck called the carotids. The clogged-pore idea is straight-out wrong, unless you're coating your scalp with sealing wax . . . or rancid hippo fat.

You can lose your hair from stress, medication, or chemotherapy, but usually the hair grows back. Genes are behind most of the bald and thinning heads out there, male and female. You can inherit baldness from your mother or your father. There's an old wives' tale, or perhaps a new one, that baldness is passed on only through the mother's side. A quick look at the countless number of bald fathers and sons will nullify this myth.

The head has about 100,000 hair follicles, little hair factories that continuously produce hair under normal conditions. When a hair falls out, a new one grows in that same follicle. Hair loss begins when a certain enzyme converts the hormone testosterone into another hormone called dihydrotestosterone (DHT), which is critical in male fetal development. The hormone also stimulates long, usually unwelcome, rigid hair on the chin and cheeks, different from a beard. Later in life, for reasons unknown, DHT starts bugging the hair follicles on the top of the scalp. These follicles continue to produce hair, but it is very fine and short, a peach fuzz. The hair follicles on the side of the head, also for reasons unknown, are not affected by DHT. They continue to produce thick hair, hence the "monk ring" of hair that is left when other hair falls out. The baldness gene (actually, there may be many, researchers think) makes too much of the enzyme that makes DHT.

That's the state of hair loss and growth today. Rancid hippo fat and goat urine are not popular remedies anymore, but equally foolish and exotic herbal remedies are readily available for purchase through the Internet. They claim to be secret formulas. I get a kick out of the implied conspiracy. "Dermatologists don't want you to know!" Yes, they do. Doctors aren't interested in holding back baldness cures; they'd make more money selling cures that work. There's no worldwide plot to keep you, your husband, or your brother bald. Closer to home, health-food stores market vitamins and minerals to promote hair growth. These are all worthless, too. Rapid, unexpected hair loss is certainly a sign of deteriorating health. By all means, see a doctor if this is happening to you. But if you are bald or your hair is thinning, your hair loss is almost certainly not a result of your diet, your circulation, your clogged pores, your poor chi, your reliance on commercial shampoos, your yin, your yang, your repugnance for goat urine, or your fondness for McDonald's hamburgers. Rather, someone in your family—maybe a generation or two away—was bald.

There is hope. Usually, hair follicles never "die" until very late in human life. Bald men and women have very tiny hairs in most of those 100,000 follicles. If the right drug comes along (and there is a very good chance one will soon), those same hair follicles can start producing longer, thicker hair. Hair transplants take hair, root at all, from the back and side of the scalp and move it up top. This works—no myth here—but the procedure can be painful and expensive. Having a skilled doctor is key to its success. Maybe you don't care about being bald. That's fine too. There's nothing wrong with wanting hair, though. Can you blame a bald person for feeling self-conscious? After all, can you name a bald U.S. president other than Eisenhower, the macho war hero?

9

ꙮ ꙮ ꙮ ꙮ ꙮ ꙮ

The Race Is Off: Race Defined

The *race* race has been run, and it looks like we all lost. Race is a social construct, deep seated in an all-too-human, "us against them" mentality; yet it was first defined only a few hundred years ago. Race, as we chose to define it, is based solely on a handful of genetic traits: skin color, hair type, and facial features. We could have picked any trait, really, to define race. We could have based race on blood type, where the intelligent and sublime ABs rule over those common, stupid, dirty Os. We could have chosen fingerprint types; there are many types, mapped out roughly from region to region around the world. We could have separated races as baritones and sopranos. If Asians ruled the world, they may have had established that Europeans were stupid because they were prone to baldness. It's all arbitrary. The ruling Europeans went with a certain set of external characteristics and tried to tie intelligence and behavioral characteristics to them.

The Human Genome Project, the NIH-fueled effort to map out the location and purpose of the tens of thousands of genes that make up human DNA, provided the definitive answer to the race question. It was the answer that biologists had suspected for the last 40 years or so—that there is no such thing as race, biologically. All modern humans descended from a tightly knit group of early humans about 100,000 to 150,000 years ago. The genes don't lie. Although humans have since spread out into relatively isolated regions of the world, there simply has not been enough time to produce the radical genetic differences that would separate

53

humans into races or breeds. From a biological standpoint, any so-called white man can be more similar to a so-called black man than to his own family. Conversely, more genetic variation exists within Caucasians than between the so-called races. Seventy-five percent of our DNA is exactly the same person to person; only the remaining 25 percent makes us different. The bulk of this genetic difference, some 85 percent, exists between people of the same ethnic group, such as Koreans. About 8 percent of genetic variations, or biological differences, will vary from ethnicity to ethnicity; and about 7 percent will vary from race to race. So, biologically, differences between Koreans and Japanese are about the same as differences between Koreans and Norwegians. Only 0.012 percent of variation in human biology can be attributed to race.

What do all these numbers mean? Let's round them off for argument's sake. Let's say that the 25 percent of genes that vary from person to person adds up to 100 genes. You could be identical to a person native to the jungles of Congo or the icy coast of Greenland except for five genes—the ones for hair type, eye color, skin color, nose size, and lip size. You share the same blood type, ear shape, right-handedness, and a cancer-protecting gene. You could very well be less similar to your cousin, from whom you might differ on ten genes: hair color, blood type, left-handedness, "detached" ears, mid-digital hair on the fingers, no anticancer protein, and so on. All you really share, obviously, is skin color. This is what is meant by genetic differences. You and a fellow native fit in one category called race yet differ in so many other respects.

Human-defined race means almost nothing to diseases such as cancer, stroke, and diabetes, the major killers in the industrialized world. Nor is race a factor in AIDS and other infectious diseases. Sure, there are different rates of diseases for different races around the world. But this is almost entirely due to environmental factors, diet, and socioeconomic status. For example, Asian Americans, after a few generations in the United States, have breast cancer rates close to the national U.S. average, which is much higher than the rates in Asia. The culprit is not genes but obesity and sedentary lifestyles. African American men in Harlem in New York City have less of a chance living to age sixty-five than men in Bangladesh, a result of the violence and social issues a black man in Harlem is exposed to. Men of African descent in Quebec, Canada, fare far better than those in Harlem.

Race (skin color) is really only a factor in skin cancer. Africans and dark-skinned people are better protected. Conversely, Scandinavians can generate vitamin D for bone development from dim sunlight better than Africans. So Africans need vitamin D supplements up north, and Scandinavians need sunblock down south. Lactose intolerance seems to haunt just about every race other than Europeans. Africans have a slight advantage when it comes to osteoporosis, but this might be because they get more sunshine, which strengthens bones indirectly. Africans carry a gene that offers a slight immunity to malaria, but then again so too do people living all around the Mediterranean, the Middle East, and southeast Asia, where malaria is prevalent. You'd be hard-pressed to show any deep difference or biological advantages against disease from one race to another.

Mentally and creatively, race is never a factor. Geniuses and idiots exist around the world. If any given race seems more inclined to behave in a certain way or excel in a particular field, the reason is purely social. Germany produced musicians; France produced painters. There's nothing genetic here. Genius is genius; society dictates the path. Rhythm is instilled in certain African cultures, hence the omnipresence of rhythm among members of the African diaspora. Europe had harmony and melody, hence the orchestrated symphony. Ireland, long treated as a second-class country by the British, produced its own blues music similar to that of African Americans, a product of being down so long.

Perhaps for practical reasons, humans are astute at finding the difference—from slight facial characteristics—between strangers who live outside the community. It took a European mind, a taxonomist to say the least, to place humans into groups like other animals. The Swedish taxonomist and botanist Carl von Linne, who went by the Latin name Linnaeus, established four categories based on external and assumed psychological characteristics in his 1758 book *Systema Naturae*. There was the European, "fair, gentle, acute, inventive, governed by laws"; the Asiatic, "sooty, severe, haughty, covetous, governed by opinions"; the African, "crafty, indolent, negligent, governed by caprice"; and the American, "copper-colored, obstinate, content, free." All the crafty, haughty, obstinate men of Europe loved these classifications. The German anthropologist Johann Blumenbach was the first to use the words *race* and *Caucasian,* way back in 1775. He divided humans into five categories, strictly on appearance: Caucasian, Mongolian, Ethiopian, American, and Malay. Blumenbach thought that the Caucasus region in modern-day Armenia and Georgia was home to "the most beautiful race of men." (Georgia *girls* were always on Paul McCartney's mind.)

Well into the twentieth century, anthropologists (usually white) were still classifying groups of people into various races and sub-races. About a hundred have been defined at one time or another. Anthropologists, however, were really only looking for a means to categorize cultures and migrations; and most never argued for race superiority. The eugenics movement in the United States at the turn of the twentieth century hoped to ban certain racial groups from entering America. Eugenicists had small successes in shutting out Chinese, Africans, and others. Most of the laws inspired by the eugenics movement were overturned by 1930 in the United States. At about the same time, the brown-eyed, brown-haired Adolf Hitler convinced Germans that foreigners were bad, despite his being born in Austria, and that the Aryan race (anthropologically speaking, from Iran), sporting blond hair and blue eyes, were superior . . . except for those blue-eyed blondes over in Poland, who would have to die. Hitler himself highlights the idiocy of attempting to define race by external characteristics.

Maybe, if groups of humans remain isolated from each other on different continents and live in isolation for another hundred thousand years, *maybe* biological races will develop, like squirrels long separated by the Grand Canyon. But who cares? It hasn't happened and it won't. The point is moot. Interracial marriages and migration have now ensured that the human species will remain one race. The eugenicists of the world will have to live with that, or stop mating with one another so that their kind will go away.

PART II

᭡᭡᭡᭡᭡᭡᭡᭡᭡

Growing Old

If you live to the age of 100 you have made it, because so few people die past the age 100.

—George Burns (1896–1996)

Back in 1967, the musical group The Who—noted in the *Guinness Book of World Records* as being the loudest band of their g-g-generation—belted out one of the most famous lines in rock 'n' roll history: "Hope I die before I get old." The drummer, Keith Moon, got his wish courtesy of a drug overdose about ten years later. The surviving band members must feel hopelessly ancient; they're all around sixty years old.

Despite popular claims, you cannot stop or reverse aging with positive thinking, hormone therapy, or youth potions. The best you can do is to maximize your chances of staying healthy through diet and exercise. Growing old, most will agree, is a drag, but it need not be *so* scary. Most of our fears of old age come from misperceptions of what it is like to be old, the ailments we assume to be inevitable. A little stiff? Must be getting old. Forget a phone number? Must be getting old. Heart problems? Only natural, you're getting old. Retirement can be a swinging time, though, especially if you make commonsense health investments early on. That is, you don't have to hope you die young.

57

10

⚕ ⚕ ⚕ ⚕ ⚕ ⚕ ⚕

Losing One's Mind:
Memory Loss and Aging

Here's a myth that even doctors believed until only a generation ago: that memory loss is simply a part of being old. Indeed, the word senility—a word that has fallen out of favor lately—derives from the Latin word for "old man." Strange. Didn't the scribes of the early English language know that the Greek philosopher and bearded old man, Socrates, remained a sharp thinker until his death in his eighties? Or that Michelangelo created his most adventurous Pietà at age eighty-nine? In the modern era, the architect Frank Lloyd Wright truly came alive later in his life, a renaissance that began when he was seventy-five and culminated with the completion of the Guggenheim in New York City, a few months after his death at age ninety-two. And the nonagenarian Milton Berle kept yapping until the end, launching a humor magazine in 1997 and a lawsuit against his old employers at NBC in 2000, before he died at ninety-three, in 2002.

The memory myth has a strong hold on our collective psyches. Memory loss is the number-one fear associated with growing old. A recent survey from the Dana Foundation found that seven out of ten adults worry that they are losing their memory, and a Bruskin-Goldring Research study revealed that 80 percent of doctors report that their patients over age thirty complain of memory loss. Most of these folks really have no difference in memory capacity from before they were thirty. They are just remembering, more often, how much they forget. A teenager who forgets the capital of Iowa

(Des Moines, I think; I should look that up) would never say he's having a "senior moment." Nevertheless, at the drop of a memory, many of us are popping gingko biloba, an herbal supplement said to aid in memory by increasing oxygen flow to the brain. Tests seem to indicate that gingko doesn't do much to boost memory, but this hasn't stopped millions of people from adding millions of dollars to the herbal market.

Mild memory loss—emphasis on the word "mild" here—is a natural part of the aging process. On average, a senior citizen can remember six items in a memory test; a thirty-year-old can remember eight items. (Hats off to the young.) Yet older people know more and have more memories by virtue of life experience. (Hats off to the old.) Aging in general is a gradual loss of vitality. Through exercise and a healthy diet and lifestyle, we control whether we lose a lot or a little of our vitality year by year. But we lose something nonetheless. Through memory training, which is essentially exercise for the mind, an old person can learn to recall thirty items. Yes, the younger person can recall up to forty items when trained, but thirty items is quite a bit better than six. Mild, natural memory loss is not debilitating, although some may feel the effects. Physicists and other scientists usually make their greatest contributions to their fields by age forty. The true scientific geniuses among us may experience a noticeable decline in cognitive ability that actually interferes with their work. Yet the vast majority of us, barring disease, will have the wits to undertake any of the mental challenges posed to us in our youth. In fact, the accumulation of life experiences *enables* writers and artists to mature and flourish far into their older years. The pioneering jazz pianists Dave Brubeck (born in 1920) and Oscar Peterson (born in 1925) sound as sharp as ever, playing in a slightly different style than in the 1950s but playing brilliantly nonetheless.

Severe memory loss—the kind that interrupts your day—is an underlying sign of disease and not a part of growing old. One memory disease is Alzheimer's, a dreadful condition that robs sufferers of a lifetime's worth of precious memories. There is no cure for Alzheimer's disease. It affects about one out of fifteen people over age sixty-five, a rate that is alarmingly high but still much lower than major killers such as cancer and heart disease. This high rate is questionable, too, because a proper diagnosis is not possible until autopsy. The cause of Alzheimer's is unknown, and the disease was never diagnosed until the twentieth century by virtue of its rarity. Today, most people in the United States live past age sixty-five, so we are seeing more and more cases of Alzheimer's. A cure would add only about nineteen days onto average life expectancy, although this should not minimize the gravity of the condition.

Alzheimer's disease is so frightening that the word has become synonymous with memory loss. Fortunately, the *main* causes of memory loss and dementia, the inability to think clearly, are treatable. The most common form of dementia is vascular dementia, a restriction of blood flow to the brain that leads to confusion. Poor nutrition is another common cause of dementia, for some seniors do not eat properly or are unable to absorb as many nutrients from food as they once did in their youth. Depression is a major cause of memory loss and confusion. Depression is far more common than Alzheimer's and presents itself with similar symptoms, sometimes leading to a misdiagnosis. Depression can lead to permanent brain damage if not treated. The same holds true for alcohol-induced dementia. All of these conditions can be minimized or reversed. These conditions are associated with aging, but they are not caused by aging.

Senility and incapacitating memory loss often befall those who no longer exercise their minds. Some folks are sharp to the very end. A French news reporter, upon interviewing Jeanne Clement, who was at least a dozen years over one hundred and the oldest human at that time, told her that he hoped to see her again in a year at her next birthday. Clement's reply: "I don't see why not; you seem healthy enough." The same went for George Burns, who outpaced other comics at his ninety-eighth birthday party.

Evidence shows that the brain develops new neurons throughout life. That is, you *can* teach an old dog new tricks. An unchallenged mind soon loses its ability to retain and utilize information. Once the body retires from work, the mind usually takes a permanent vacation as well. There are no more details to "worry" about. No more planning out the week's events, no more mentally preparing for the day's work. And the situation can turn worse. Many elderly people live in isolation, comforted only by the mind-numbing glow of television. They cannot read for failing eyes; they cannot socialize for lack of transportation. Just as our bodies need exercise to stay healthy, our minds need new stimuli in order to work properly. Even a young person placed in this situation will lose the ability to think clearly.

An old mind that is stimulated will create new networks for storing and receiving information, just like a younger mind. We maintain an unlimited capacity to learn. Many people are surprised by this fact because, frankly, we see so few examples in our daily lives. We see old folks retiring physically and mentally. The issue is social, not biological. Occasionally, we hear a human-interest story concerning some ninety-year-old graduating from college. Recently from Peru came news of a 102-year-old woman who signed up for that nation's new literacy program; it had been this woman's life-long goal to learn how to read. These are examples of extraordinary individuals doing something quite average. We all have this capacity. A great person, however, will learn for learning's sake.

Learn a new language. Learn another new language ten years later. Balance your checkbook without a calculator. Draw a complicated pattern with your right hand and then try it with your left hand, and keep on practicing until you get good with your left hand. Learn to play pinochle, bridge, or another memory-intense card game (the more social the better). Write your life story. The list of possibilities goes on and on. What's the use of learning a new language every ten years after age sixty-five? This won't get you a pay raise in your retirement check; it won't get you a new job as a translator; it might not even get you to the country whose language you are studying. Learning a language—or musical instrument or any of a host of skills—will solely buy you years in terms of a healthy mind. You won't get smarter, per se, but you can minimize natural memory loss. The healthiest, oldest old have this mentality. They continue to learn without thinking twice about it; they continue to form new neurons throughout life. And they are rewarded with healthy, sharp minds.

Let's not forget wisdom. Some memories may fade, but wisdom can only be gained through years of experience. As Victor Hugo wrote: "If you look in the eyes of the young you see a flame. If you look in the eyes of the old you see a light."

‖

₹ ₹ ₹ ₹ ₹ ₹

Getting Stiffed: Vitality and Aging

The folk guitarist Doc Watson is still one the best "finger pickers" around today. Born in 1923, he can no longer play as long as he used to, but there is no denying his proficiency. I saw him in concert in 2001. Aging, it seems, has done nothing to compromise his guitar playing other than cutting back on a few concert dates.

Aging involves a gradual wearing down of body parts. You cannot stop it and you cannot reverse it. No one is too happy about this. The aging process, however, need not interfere with everyday life. Healthy old people walk, shop, cook, and clean. They cannot play professional baseball, but they can play baseball nonetheless. As for Doc Watson, he can no longer play several hours a night, three hundred days a year, but he can manage a hundred concert dates playing for ninety minutes. Not too shabby.

When it comes to aging, there seem to be two camps: those who believe that aging can be reversed and that stiffness and frailty are your own fault for not thinking "young" or exercising, and those who assume we are all destined to shrivel up and break in half when we get old. The former is just plain silly. All animals slow down as they age. Dogs and cats, two of the few animals in the world protected from predators and starvation and actually allowed to grow old, age in nearly the exact same way as humans, regardless of their preconceived notions of what it means to be an old dog or cat. As for the latter notion, Doc Watson is proof positive that old age need not be debilitating.

The machinery of antiaging business and research is humming now that the baby boomers, some with considerable wealth and political clout, have reached age fifty. Yet even with recent scientific advances, your chances of expanding your life span are slim . . . unless you're a parasitic worm or a fruit fly. Humans, who for the most part are more complicated than these creatures, can reap little benefit from the antiaging potions and precepts now flooding the marketplace, which are at best naively optimistic and at worst fraudulent and harmful. Credible scientists do agree that promising medical research may someday lead to methods to slow the physical degeneration associated with aging and even extend the human life span, which for now appears set at about 120 years. But, despite claims to the contrary by those selling products and regimens, science hasn't done anything like that yet. For now, every book, powder, or pill that promises a fountain of youth—skin that doesn't age, organs that keep putting out, an immune system that never weakens, a sex drive that never droops—is just plain wrong: misguided, excessively hopeful, or outright deceptive. This includes hormone treatment, antioxidants, and all those seminars on thinking "happy thoughts." Clearing a wrinkle is not tantamount to growing younger. Age reversal is merely a lucrative euphemism for losing weight and getting in shape.

The best you can hope to do is stay as healthy as you can for as long as you can. Aging happens. Three physical conditions associated with aging, but not caused by aging, are stiffness, frailty, and a loss of libido. Exercise and a healthy diet, the same old mantra, will help minimize this trio of misery.

Vitality is less about old age and more about being in shape. A healthy seventy-year-old is no different than an unhealthy thirty-year-old, biologically. An unhealthy seventy-year-old is, well, in pretty bad shape. This is true even when it comes to sex. Smokers, the obese, or otherwise out-of-shape men in their thirties will have more difficulty achieving and sustaining an erection than a healthy seventy-year-old man. In fact, of the some thirty million men in America encountering some degree of erectile dysfunction—a kinder, more precise term for impotence—about half are in their forties and fifties, often with diabetes or circulatory disease. For women

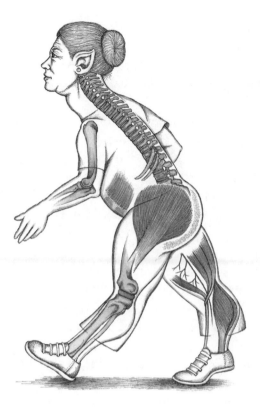

Medicine may allow us to live past 100 years, but is the human body prepared? We will need a major makeover: backwards-bending knees to alleviate deterioration; added ribs to better support organs; forward stance to alleviate pressure on the vertebrae; thicker padding between vertebrae and joints; shorter stature to prevent falls; thicker bones; larger ears to collect sound with better efficiency; sharper, more durable eyes; veins and arteries with more check valves to better pump blood and clear pathways of fat droplets; and much more.

Illustration by Patricia Wynne

too, exercise ensures better blood flow to the genital region and thus better sexual satisfaction. (A woman's diminished sexual activity later in life is often due to a lack of a partner, not poor health.)

How do you avoid being an unhealthy seventy-year-old? Walter Bortz, an aging expert at the Stanford University Medical School, compares aging to an athlete slowly losing his edge after peaking. Bortz found that athletes' performance declines half a percent a year as they age, assuming that they remain as fit as possible. If you stay fit, you can keep 90 percent of your vitality by age 70. That's not bad. If you don't stay fit, like some athletes, you will lose 2 percent or more of your vitality each year and have only 30 percent of life's vitality by age 70. Bortz admits the math is a bit fuzzy, but his vitality loss concept is an instructive analogy nonetheless. Those folks in their sixties and seventies who seem so feeble, the same people fueling stereotypes of the elderly and our misconceptions of what it means to be old, have lost most of their vitality, à la Archie Bunker, Bortz says. Vitality trickles away from years of poor diet, inactivity, stress, and depression. As with a decline in memory and learning, this is a social issue, not a biological one.

You'll never regain the vitality of your youth, but, at age forty with less than optimal health, you can regain some of the vitality of a healthy forty-year-old with proper exercise and diet. This is not age reversal. You are merely recuperating from loss due to inactivity. Imagine a leg in a cast for two months, shriveled from lack of activity. Exercising brings that leg back to normal. The leg didn't get younger. Stiffness can come (and go away) at any age. Frailty too, diminishes with exercise and a calcium-rich diet at any age.

Arthritis and osteoporosis are diseases, not a natural part of growing old. Osteoporosis involves runaway leaching of calcium from the bones, often preventable and sustainable with calcium supplements and exercise. Arthritis is a gradual wearing away of cartilage, the protective material that covers and cushions the ends of bones. With cartilage gone, bone rubs against bone, which feels truly miserable. Arthritis has its obvious causes—repetitive and unnatural movement, sports injuries, slovenly posture, or obesity adding excess pressure on joints—yet the disease is often genetic. You can considerably lower your risk of developing arthritis, and minimize the symptoms if you have arthritis, by exercising and drinking up to two quarts of water a day, according to studies conducted by the National Institutes of Health.

As relayed in Part IV of this book, antioxidants are not the youth potions they are purported to be. Injections of human growth hormone or DHEA come with a litany of side effects, and at best produce only the "antiaging" results that exercise can bring. Expensive seminars on "thinking young" do nothing other than encourage you to exercise, which is just plain practical, not youthful per se. Drinking water, walking, stretching: the true elixirs in life are cheap. These will not only keep you from getting stiff, they'll also keep you from getting stiffed—by antiaging hype.

12

ও ও ও ও ও ও ও

Illness Gets Old: Aging and Disease

There's an old joke that the trick to living to one hundred is to get to ninety-nine. This isn't so far off. Once you get past eighty, your odds for getting cancer go way down. Few people over eighty-five have heart disease. All the major health hurdles are cleared by age ninety. You've won the race at this point. Aging doesn't cause disease; a lifetime of bodily abuse causes disease. Take care of yourself, and you'll have a good chance of experiencing disease-free aging.

Thomas Perls of Harvard Medical School is conducting a study on centenarians. Over 30 percent of his centenarian patients claim to be in excellent health, feeling the way they did in their seventies. Over 40 percent report good health and 20 percent report fair health. Only 2 percent are frail. Most do not have a family doctor, nor do they feel they need one. Perls thinks that all of us have the genes to reach our nineties; the centenarians, though, might have a slight genetic advantage in avoiding disease early on.

Aging is a natural part of life, not a disease itself. Some people may be more susceptible to the diseases and conditions that prevent them from reaching old age. The big three are cardiovascular disease, cancer, and stroke. These all strike largely between ages thirty and sixty, and they are loosely associated with genetics. So, even if one of your parents died young from disease, don't assume that you are also doomed to die of that same disease. While it is true that certain genetic disorders nearly guarantee certain types of cancers or conditions in adulthood, most often it is lifestyle (diet, exercise, exposure to toxins) that brings about diseases.

The big abuses that prevent most people from reaching old age are cigarette smoking, alcohol abuse, reckless activity such as sky-diving, exposure to harmful chemicals at the workplace, years of excessively hard labor, and obesity. If you can avoid these, you have considerably increased your chance of living healthily into your nineties or hundreds. Other factors include exercise and diet. Daily exercise lowers the pulse rate, strengthens the lungs, and increases blood circulation, major factors in warding off certain diseases. A diet rich in vegetables, fiber, low-fat foods, and nonan-imal protein also considerably lowers the risk of diseases, such as the big three mentioned above. A variety of vegetables ensures that the body is properly nourished with necessary vitamins and miner-als. Fiber strengthens cells and aids in the proper elimination of food. Low-fat foods ensure that the blood passageways remain clear of fatty deposits that can lead to heart attacks and strokes. High amounts of animal protein (from beef and pork) act to leach calcium from the bones, making them brittle. Some meat is fine, though; and vegetarians do not live longer than light meat eaters.

Presbycusis refers to the loss of hearing in old age. The tiny bones in the middle ear that vibrate with incoming sound waves begin to noticeably fail after age sixty. The ability to hear high-frequency tones may decrease by as early as age thirty; low-frequency tones by age sixty. Men lose hearing more than twice as quickly as women do. All of this doesn't mean you'll be stone deaf, though. Hearing loss is inevitable, but you can protect what you've got. The most important point is to avoid loud noises. Hearing experts say that the younger generation, from kids to young adults, will have major hearing problems by the time they are fifty years old. The culprit is noise—noise from headphones, noise from bars and music concerts, noise from the countless cars, lawn-mowers, leafblowers, and other power equipment ubiquitous in modern, industrialized society. The evidence is in. Already doctors have recorded that the majority of men in their forties have the hearing of a sixty-year-old. Pete Townsend of The Who, who hoped he'd die before he got old, now wears two hearing aids, as do countless other musicians who have taken the stage after the birth of the amplification era.

The eyes take a beating, too. Nearly everyone will encounter difficulty focusing on close objects by age forty and difficulty reading the fine print of a newspaper by age seventy. Fortunately, the eye conditions that truly affect life quality are nearly all preventable. Good health matters greatly. As the incidence of diabetes in America continues to rise, so do the rates of glaucoma and diabetic retinopathy, diseases that can cause blindness early in life if left untreated. High blood pressure can also put strain on the eyes, weakening tiny blood vessels in the retina. As with hearing, truly debilitating vision is associated with a lifetime of poor health, not aging.

Dying of old age just may be the best way to go. If you have lived to 100, this means you were more or less healthy your entire life. Yes, 100 seems magical. Even more magical to me is 114.16 years; that translates to one million hours lived.

13

֍ ֍ ֍ ֍ ֍ ֍

See You in 2150: The Long and Short of Life Span

odern medicine sure is great, isn't it? We've wiped out
nasty diseases. Organ and limb transplants are becoming
commonplace. Cancers can be removed; once-terminal conditions
can be reversed. Not a month goes by without some remarkable
health development making the headlines. Yet, as far as we have
come, has the human race increased its life span? Not at all. This
is one of the biggest misconceptions about old age: we are not living any longer. Human life span, unchanged for the past 100,000
years, has remained fixed at a maximum of about 120 years.

The confusion lies in the term "life expectancy," the average age
we as a nation live in any given era. I myself "expect" to live much
longer than the national life expectancy. Today, in the United
States, life expectancy is 72 years. In 1900, life expectancy was 47
years. In 1776, when the country was founded, it was way down
at 35 years. In Roman times, humans lived an average of only 25
years. The numbers can be deceiving, though. The Greek philosopher Socrates died just before his 90th birthday and that was by
execution. Several early Christian priests in the fourth century
lived into their 90s, and Saint Anthony was 105. Michelangelo
was chiseling a Pietà at age 89. Ben Franklin died at 84. Several
Native American chiefs lived into their 100s, including Chief

Joseph and Chief Red Cloud (who broke his legs in a car crash at 100 and died of pneumonia at 111).

The life expectancy number is an average, taking into account all the babies who die before their first year, all the young men who fall in war, and all the souls who die from diseases along the way. The Romans loved a good battle, hence the paltry life expectancy. In early America, the infant mortality rate was quite high; one in nine children never saw a first birthday. In 1900, kids and adults alike were dying of the diseases we have since brought under control or licked altogether: measles, polio, smallpox, dysentery, and water-borne diseases. These factors added up to an overall life expectancy unimaginably low by today's standards. Romans didn't die at age twenty-five. They may have lived to age ninety or eighty or twenty or two. Average all these ages and you get about twenty-five.

Both the highest infant mortality rates and the lowest life expectancy rates in the world are in central Africa. In some African regions ripped by the AIDS epidemic, the life expectancy is as low as twenty-five years. It is not uncommon to hear a news report, however, of some local woman in Uganda passing the hundred-year mark. She somehow managed to avoid war and beat the diseases that took her compatriots. Modern medicine most likely did not come into play.

Perhaps a more revealing longevity statistic is how long people live once they reach adulthood. In the United States, a male baby's life expectancy is seventy-two years. If he lives until thirty-five, his death supposedly won't come until he's seventy-eight. If he lives to sixty-five, then he can expect to live almost two decades more, until eighty-two. From country to country, life expectancy evens out as a person reaches adulthood and middle age. That is, nearly everyone worldwide over sixty-five has a good chance of living another ten years or so. Overall, Japan leads the life expectancy contest, followed closely by Iceland, France, Switzerland, and Germany. In Japan, the island of Okinawa boasts the highest percentage of centenarians, a result attributed to active lifestyles and a low-calorie diet of mostly vegetables, rice, sea products, and the occasional slice of pig cheek.

Some scientists say that no one before the year 1800 lived longer than one hundred years. A thousand years ago, they say, no one lived past fifty. This simply isn't true. Aside from the well-documented lives of aged historical figures, there were no doubt countless "ordinary" people—farmers, carpenters, sailors—who lived to see their great-great-grandchildren. In fact, there are folks today living on the Italian island of Sardinia and in the mountainous regions of the Caucasus who live well into their hundreds. If these peoples can live this long without the benefit of modern medicine, then their ancestors likely shared this longevity a thousand years ago. These cultures' lifestyles haven't changed much in that time. There are parts of the Caucasus where the number of centenarians is 1 in 140, compared to 1 in 5,099 in the United States. (Admittedly, reports can be exaggerated. Those old folks in the Dannon yogurt commercials apparently lied about their age, either while in their teens to avoid fighting in Russian wars, or later in life when adding twenty years to your age made you a celebrity.)

So how old can we live? In the Book of Genesis, Adam and his descendants lived hundreds of years. Methuselah was the fittest, reaching 969. Clearly, this is folklore. But it is interesting to note that God in Genesis, tired of wicked humans living so long, set the new age limit at 120 years. This number happens to be about the oldest documented age of any human in modern times. A few women have actually made it to 120; the late Jeanne Clement of France holds the world record at 122. Quoting the Bible once more, we find King David pining about life in Psalm 90: "Seventy years is all we have, eighty if we are strong." Not 120, but not bad for a world of war and pestilence—and certainly a strong indication that we haven't improved much upon our life span.

Longevity experts such as Dr. Walter Bortz of Stanford University Medical School say that most humans will live over one hundred years later in the twenty-first century. Dr. Thomas Perls of the Harvard School of Public Health is a bit more conservative, stating that the average person has the genes to get him to eighty-five and then maybe ninety-five with good behavior. Then there's Dr. Michael Rose of University of California, Irvine, who predicts that genetic tampering can get us to three hundred. There's no evidence to support Rose's claim, but it's nice to think about nonetheless—particularly if those three hundred years are happy ones. (The social impact of seventy-five-year-olds still bumming around Europe after college will be challenging indeed.)

Basic sanitation and access to clean water at the turn of the twentieth century in developed countries has had the biggest impact on life expectancy, pushing it up to about age fifty (from age thirty-five through most of history). Japan's rise to the top of the life expectancy ladder came after World War II with a marked decrease in infant mortality. Midcentury, antibiotics and vaccines added another fifteen years to life expectancy. Surgery and medicines from 1970 to the present added another ten years or so. The elimination of the big killers—heart disease, cancer, and stroke—will add another fifteen. This would get us to about age ninety-five.

A revolution in the science of antiaging is needed to improve on a generous life span of 120 years. One promising path would be to somehow slow the aging process, perhaps through genetics or caloric restriction, techniques that have increased the life span of rodents. Humans aren't rodents, though, and scientists say we are years from understanding how and why we age, let alone understanding how to control aging. For now, no elixir, including popular hormone therapies, will increase human life span. While hormone therapy may increase strength and stamina, the side effects are unknown, and exercise works better than hormone therapy, anyway. All so-called antiaging medicine—from crushed deer antlers and antioxidants to human growth hormone, all readily available over the counter—is bad medicine. Happy thoughts and positive thinking, as purported in popular health books and alternative medicine circles, won't slow aging either. If staying young is simply a state of mind, then animals wouldn't show the signs of old age. Domesticated dogs and cats age in nearly the same way as humans.

If it is any comfort to the matchmakers out there, humans are closing the aging gender gap. A quick glance at nursing home populations would lead you to believe that women live longer than men. This phenomenon may be more social than biological. As more and more women enter the workforce and assume roles traditionally held by men, they surrender their cherished status as longevity leaders. In the United States, the life expectancy rates among men and women have narrowed considerably over the last twenty years. In some developing countries, men already easily outlive women (who die during childbirth or from a lifetime of hard labor). Demographers suspect that life expectancy rates among women and men in industrialized countries will level out in the twenty-first century. This equality of longevity is apparent today among the very old. In the United States, men make up 20 percent of those over 100 years old but 45 percent of those over 105.

Now, to rephrase an immortal Mark Twain quip, the reports of our longevity may be greatly exaggerated. While many doctors agree that more and more people will live into their hundreds in the coming decades, the overall life expectancy for the United States and other industrialized nations may have peaked and may even be heading down. The worry is that most young people today are simply not healthy—they're too well fed, too sedentary, and thus too overweight to escape the clutches of diabetes, cardiovascular disease, cancer, and stroke. These are the top killers in developed countries, and more and more people are at high risk of dying from these diseases by virtue of their expanding waistlines.

As it is, the United States is not exactly leading the pack when it comes to life expectancy rate. American children born in 1996 have a life expectancy of seventy-two years, according to the U.S. Census Bureau's International Programs Center. This is at least two years less than life expectancy in Japan, Singapore, Canada, Israel, and most of Europe. Russia, however, fares far worse than the United States, with a life expectancy for men at around fifty-nine years. Interestingly, Russia's life expectancy rate rivaled the U.S. rate through the 1960s. What a difference a generation makes. The leading cause of death in Russia is cardiovascular disease, followed closely by accidents and violence. Vodka is often the hidden culprit here, a major contributor to the shortened life expectancy rate. The average Russian consumes 4.4 gallons of alcohol a year, the highest level worldwide. Russia's crumbling economy and loss of world status has exacerbated the problem. Will America's life expectancy rate drop as low as Russia's rate? Are fatty foods and a lack of physical activity the equivalent to Russia's taste for vodka? There is no need to paint a picture of doom for the United States, but what is certain is that an above-average life expectancy rate can change dramatically for better or worse in a single generation.

14

On and On: Longevity and Genetics

American baby boomers are getting old. Hair is going gray; wrinkles are forming around eyes. Sales of antiaging potions and health books are at an all-time high. These folks are desperate, and they have the money and political savvy to influence nationally funded medical research. So the search for the longevity gene, however futile, is under way in laboratories across the United States.

Does such a gene exist? No one is sure. Some centenarians living today have siblings in their nineties. Many more do not. So it is certainly not obvious that there is a genetic connection for living long. Many scientists believe there may be a small group of genes that regulate the speed at which a person ages. This is hotly debated, but the so-called slow-aging people would be those who look great at eighty, run marathons at ninety, and play golf at age one hundred. They, the theory goes, have this set of longevity genes that may exist on chromosome 4. Clearly our genes play some role in keeping us alive: some individuals are better equipped, genetically, to fight off cancer and heart disease. Such individuals, naturally, would have a better chance of reaching an old age.

Don't get too hung up on not having the right genes, though. According to the noted aging expert Leonard Hayflick and many others, longevity is much more a factor of nurture than of nature. In other words, for most people, exercise and a healthy diet—more so than your genetic blueprint—will set you on the path to a long life. Hayflick argues that genes have no direct role in aging because humans don't age at the same rate after reaching adulthood. In contrast, they age almost identically from infancy to sexual maturity, hitting puberty and various cognitive landmarks at nearly the same age. Genes seem to dictate biological development up to about age twenty-five but provide no further instruction for aging. Also, Hayflick argues, genes that promote slow aging would have no way of being selected, or passed from one generation to another, in evolutionary terms. The body is only interested in getting you to reproduce and then take care of your children. Genes that enable a person to live to a hundred offer no advantage over genes that enable a person to reproduce by age thirty, so evolution would have no reason to favor the slow-aging genes.

Thomas Perls of Harvard Medical School, the founder of the New England Centenarian Study, is one scientist who is looking for longevity genes. In 1997 Perls started the Centenarian Sibling Pair Study, recruiting centenarians and their siblings of ninety-plus years from around the world with hopes of finding genetic similarities. His work builds upon the famous Danish Twin Study, which showed that only 30 percent of longevity was due to genetic factors. However, folks in the Danish study were living only eighty-some years. Perls is studying very old people, in their nineties and hundreds. It is with these very old, the "superstars" and "Michael Jordans of aging," as Perls calls them, that we may see hints of genetic influence. In August 2001, Perls announced significant progress in the search for these genes. He and his colleagues have since established a commercial venture called Centragenetix to identify the genes and develop therapies that might give all people the same advantage that centenarians purportedly have in fighting disease. The name of the company is so convolutedly futuristic that it is bound to be successful.

To understand how genes could possibly slow aging—not just fight cancer and heart disease but actually allow people to live longer—consider the maximum number of times a cell can divide. Hayflick was a freshman researcher at Philadelphia's Wistar Institute in the 1950s when he noticed that cells cultivated in a test tube only divided a set number of times before the entire culture was dead. He eventually found that human cells in a dish divide consistently about fifty times. Then they slow down and stop dividing completely. This maximum number of divisions is now called the Hayflick limit. The cells of mice have a Hayflick limit of about thirty. Mice live only a few years. Human cells have a Hayflick limit of about seventy. The maximum recorded life span is 122 years. What tells the cell to stop dividing? There appears to be a little cap on all of our chromosomes called a telomere. When the cell divides, the chromosomes are duplicated. The resulting pair still has a telomere cap, but now the cap is a little shorter. After another cell division, the cap becomes smaller yet. After dozens of cell divisions, the telomere cap is nearly nonexistent. It is at this point that the cell slows down, stops dividing, and dies. Those who live over 100, some scientists believe, may have a special gene that helps rebuild the telomere. The jury is still out on this theory. Until then, doctors know of nothing we can eat or do that will increase our Hayflick limit. Even if we could, there's no proof that extending cell life will translate into prolonged overall longevity.

One common characteristic among the very old that is clear to Perls and other aging researchers is not their genetic makeup or telomere size but rather their attitude. Centenarians have a zest for life. Nearly all have a very active lifestyles, a virtual nonretirement of engagements and house chores and even speaking tours. And nearly all have had healthy lives from the start, eating only in moderation and exercising naturally with the daily tasks of life: walking or biking to work, walking up stairs, choosing manual and mental labor over electrical gadgetry. Virtually no one living over 100 was ever obese. Yes, a few seem indestructible; smoking and drinking doesn't affect their lives. (The tobacco industry actually tried to hire one 105-year-old lifelong cigar smoker from Denmark as a spokesman, but he declined. I don't want to imagine the pitch: "Tobacco: It's doesn't kill everyone!") But attitude may be the true fountain of youth. This is not mind-body healing or happy thoughts. Rather, the general attitude among centenarians provides for unconscious yet prudent health choices throughout life. They think it is natural to stay healthy and active; thus they naturally do the things that keep them healthy and active.

PART III

Enough to Make
You Sick

Be careful about reading health books. You may die of a
misprint.

—Mark Twain (1835–1910)

After several hundred thousand years of cluelessness, humans finally
figured out the nature of most diseases by the late nineteenth century. We're still generally clueless, but we're getting better. We know
that viruses and bacteria are the root causes of most communicable
diseases. A mysterious, newly discovered, quasi–life form called a
prion is behind Creutzfeldt-Jakobs (mad cow) and other brain diseases. Radiation and certain chemicals can cause mutations, or
changes, in human DNA that cause the body to do foolish things,
such as making cancer cells. This progress isn't bad for a century's
work following the establishment of germ theory. Nevertheless,
our confusion lingers over the existence, cause, and cure of certain
diseases.

15

The Plague Lives! The Black Plague in the Modern Age

The Black Plague did not go the way of the lutes and mead of the Middle Ages. Today, worldwide, there are up to several thousand cases of plague each year, including an annual average of about twenty in the United States, according to the U.S. Centers for Disease Control and Prevention (CDC). The plague is one of three diseases subject to the international health regulations of the World Health Organization. An outbreak affecting tens or hundreds of thousands can strike without warning today. We have not eliminated the potential for a plague outbreak. We have merely improved upon containment and survival rate. About 85 percent of those infected will survive, thanks to antibiotic drugs.

The infamous Black Plague, most prominent between 1347 and 1352, killed just about everyone in Europe who contracted the disease—about 25 million, or a third of the population all told. This outbreak actually started in Mongolia and spread to trading ports in China around 1330, killing 30 million in Asia before even reaching Europe. Rats on trading vessels spread the disease to the seaports of Italy, where it then traveled on land throughout Europe. This plague outbreak wasn't the first. The disease known as the plague has most likely existed since the dawn of humankind, and one particular outbreak soon after the collapse of the Roman Empire was equally as devastating as the medieval plague. Rome, in fact, suffered though at least ten major bouts of the plague during its glorious empire days.

The Bubonic Plague. The Great Plague. The Black Plague. The Black Death. Many names exist for this nasty disease that kills within one week of infection. Victims often suffer high fevers, delirium, and swollen lymph nodes (called buboes), which remain extremely tender and frequently pop and ooze through open sores. Black spots cover the body, and fingers and toes turn black from gangrene—tissue death from a lack of blood flow. The three major forms of the disease affect slightly different regions of the body, yet each is caused by one bacterium: the *Yersinia pestis*. This bacterium is spread by fleas living on rodents, most commonly on rats. For all of history, the plague has really been the rat's burden; plague bacteria survive from one generation to the next in the belly of fleas that live on these rodents. Occasionally, infected fleas bite humans or humans' cherished pets, the cat and the dog. This can happen when rodents live in close contact with humans and, particularly, when there is a massive rat die-off and the fleas scurry for another type of animal host.

Dogs seem to do all right; their immune system can fight the disease. Cats and humans aren't so lucky. A human can contract the plague from being bitten directly by an infected flea; from handling a dead, infected animal (such as a hunter cleaning a freshly killed squirrel or rabbit); from inhaling infected droplets from a coughing cat; or from the bodily fluids of another human. The fourth scenario is actually the least likely; there has not been a human-to-human transmission of the plague in the developed world since 1924, according to the CDC. However, human-to-human transmissions were likely quite significant during the major plague pandemics.

The first known pandemic—that is, an epidemic that circles the world—started in Central Africa and killed tens of millions throughout the Mediterranean and beyond in the sixth century. The medieval plague was most deadly during its first five years, but some historians argue that the same plague lasted for hundreds of years after this, wiping out entire villages without notice. The most recent plague pandemic started in northern China and hit Hong Kong and Canton by 1894. The plague quickly spread to other continents through these active seaport regions and killed several million people by the end of the century, including six million in India alone. Insidiously, this recent plague brought the *Yersinia pestis* bacterium to parts of Africa and the Americas previously considered plague-free. The bacterium settled in these local flea/rodent populations and continues to be the source of the minor outbreaks we encounter today.

Scientists believe that the plague bacteria have disappeared, for now, from Europe and Australia. In North America, the plague bacteria are present in fleas on rodents—rabbits, squirrels, and mice, mostly—throughout the Pacific Northwest and the Great Plains, and as far north as British Columbia and Alberta. The bulk of human cases occur around the Native American reservations of northern New Mexico and northern Arizona. Los Angeles was host to the last urban outbreak in the United States, in 1924, when thirty-one of the thirty-three infected people died. The city was quarantined to control the spread.

Is the Great Plague just resting? Can the plague rise from its sleep and once again kill millions of humans? Yes, particularly in the crowded cities of developing countries. The problem is that the plague acts fast, killing in a week. There is treatment, but a person has to receive the medicine quickly. There is also a preventive vaccine, but its effectiveness has not been conclusively demonstrated. Also, the vaccine may not be readily available in large quantities for distribution. If developed countries are slow to react to an outbreak in a developing country, millions could die. The 1994 India outbreak, the last large epidemic, was fortunately confined to less than 300 deaths. This plague started with a burst in the local rat population dining on free food brought by disaster relief crews of the great 1993 earthquake, which killed 10,000 people.

The plague, in many ways, is a rat disease, and two scientists from Cambridge University are looking at the situation from the rat's point of view. Drs. Matthew Keeling and Chris Gilligan have developed a model of why the plague seems to lie dormant for decades or even centuries. Take London, for example: over 10,000 people died of the plague around 1590; then things were quiet for 15 years with hardly any deaths; then 30,000 deaths around 1605; then quiet; then 30,000 more deaths around 1625; then quiet again;

The Black Plague never went away. Health workers examine dead rats (without gloves!) during a plague outbreak in New Orleans at the turn of the twentieth century. *Courtesy of the National Library of Medicine*

and then 10,000 deaths approaching 1640. The disease (that is, the bacterium) continues to live regardless of whether humans are dying.

The plague has plagued rats for millennia, killing them at times in large numbers. During rat plague epidemics, over 95 percent of the little guys might die. The few survivors develop a resistance to the disease. Fleas carrying the plague bacteria feed contently on these rats, and these rats don't die. The rats mate, and many of their offspring will have a natural resistance to the disease. If the occasional rat doesn't have resistance, it will die, and its fleas will hop over to another rat. Over the years, though, fewer and fewer rats will have resistance to the plague, which itself is mutating to ensure its survival. More and more rats die. The epidemic hits, wiping out all those rats that aren't resistant. The fleas are hungry, though. They need somewhere to feed. Without living rats to feed on, they hop onto cats, dogs, and humans. Now the epidemic has hit the human population. Killing the rats won't work, because that just sends more fleas more quickly to another animal host. Keeling and Gilligan suggest keeping the rat population consistently low. Overcontrolling the rat population during a human outbreak will only hasten the epidemic. The CDC advocates killing fleas, not rats, and recommends flea collars for dogs and cats.

The Italian writer Giovanni Boccaccio, who lived during the medieval plague, wrote that the plague's victims "ate lunch with their friends and dinner with their ancestors in paradise." This is a testament to the speed and severity with which the plague strikes. To our advantage today, we at least know what causes the plague. The scientists who lived though the three major pandemics did not—and this may be the reason why the outbreak was global. Knowing what *Yersinia pestis* looks like greatly eases the burden of controlling it. The World Health Organization (WHO) remains vigilant about global surveillance of the plague. Any confirmed case of the plague must be reported to it within twenty-four hours. The WHO oversees strict quarantine measures for travelers to and from plague-inflicted regions.

Closer to home, it is true that anyone in the southwest United States can easily contract the plague. The Black Plague is real. It is alive and affects us today. Should you worry about it? Probably not, unless you're the type who goes camping in New Mexico, comes in contact with rodents, starts getting sick, and decides to put off going to the doctor for a week. Is the Black Plague the worst disease that has hit the earth? Maybe not. At its worst, the plague has killed five million humans in a year. The Spanish Flu of 1918 killed twenty-five million. And a flu pandemic is far more likely to recur than the Black Plague.

16

ಕ್ಕ ಕ್ಕ ಕ್ಕ ಕ್ಕ ಕ್ಕ ಕ್ಕ ಕ್ಕ

Cold Comfort:
How to Catch a Cold

Got a tough Chicago cold? Need a cold medicine worthy of those rugged winter warriors unfortunate enough to live in Green Bay and Buffalo? Cold medicine commercials pump up the myth that people in the coldest cities endure the nastiest colds. Leave home without a hat or scarf in the wintertime, and you're certain to come down with a cold and fever. Get your feet wet, and you'll run the risk of pneumonia. Not true. Warmth-loving viruses, not drafts and blizzards, cause colds and pneumonia.

There are over two hundred types of cold viruses and dozens of strains that induce a respiratory pneumonia causing a billion cases annually in the United States. Different viruses attack different parts of the body. This is why there are head colds and chest colds. The viruses, ten to fifty times smaller than bacteria, are such simple organisms that they need animal cells for everything: food, shelter, and reproduction. They contain only about ten genes. They invade the human body, enter into cells, borrow cellular material to multiply, and, in the case of colds, wait to be sneezed out so they can conquer someone else.

So what's the seemingly obvious connection between colds and the cold? Wintertime is a season when everyone stays inside, huddled together with the windows closed. Close quarters, no fresh air: viruses easily spread from person to person. Making matters worse, the cold virus life cycle is most active and more prevalent in the wintertime. That's just the way nature works. Mosquitoes are active in summer; cold viruses are active in winter. If the cold virus were more prevalent during the summer, maybe we would have associated them with that weather and called them "hots." Viruses don't even like cold weather. That's why they seek our warm bodies. Cold viruses multiply best around 33 degrees Celsius (91°F), about the temperature in the human nose. Left exposed—on a doorknob or countertop—they die in a few hours.

So, the virus goes around, and you have little chance to escape it because you cannot open a window to let some fresh air in, or get out of the house all together. Still, you might think that you would catch a cold after being caught in a piercing, bone-chilling rain. Not necessarily. If there is no virus around, you won't catch a cold or pneumonia, no matter how soaked you are. That fever and possible nauseated feeling you experience after a good soaking is just your body trying to regulate your temperature, kicking into overdrive to balance out that noxious external exposure. The fever will quickly pass once you warm up. The same is true about your runny nose. Your body's immune system feels inundated by the cold weather and is building up a defense to battle possible invaders.

But still, you say, you've read *Wuthering Heights* and you know all about life on the brutal moors. People catch colds and pneumonia from exposure to the cold and damp, don't they? Lord, you're pesky. All right, I'll tell you why.

Cold weather compromises your immune system, your body's defense against viral and bacterial invaders. When your body is warm and at rest, it manufactures white blood cells and other cells of the immune system to fight potential diseases. When your body has to work double-time to heat your body in subfreezing weather, it is not making immune system cells. Instead, in the process of warming you, your body lets its guard down just a bit. Resources are diverted. So when you are physically cold and consequently exposed to a virus, your body might not be able to disable that virus as well as it would have if you were warm and comfortable. The virus can gain an upper hand, multiply in your body, and cause the symptoms commonly known as "having a cold." Stress in general compromises the immune system. Stress can come from a lack of sleep, from working too hard, from exercising too hard, from being too cold *or* too hot, or from being in tense situations at work or at home. When you feel run down for any of these reasons, your body is particularly vulnerable to catching a cold. Being cold and wet is just one type of stress.

Cold weather does affect the human body in one unique way: it numbs the cilia in the respiratory tract. Cilia are fine, hairlike fibers that filter pollutants and expel foreign material from the lungs, such as viruses. When the cold weather chills the cilia, viruses can more easily enter into lungs and ultimately the bloodstream. (Smoking also numbs the cilia, and this is why smokers have more colds than nonsmokers.) Again, all these factors merely help the virus get the upper hand in the body. If no virus is present, even if you are really run down, you won't catch a cold. Tired, cramped, stressed, and cold scientists in Antarctica and in the Arctic rarely have colds because there are few people around to spread a cold virus.

This myth-busting, however, shouldn't give you an excuse to head out in the cold without dressing warmly. Hypothermia can kill; frostbite can lead to loss of fingers and toes. Hypothermia occurs when your body temperature drops below the normal 98.6 degrees. You know those nuts who dance around drunk and barechested at football games? Many of them end up at the hospital with a core body temperature of about 90 degrees. If the body stays that cold for more than an hour or so, it will shut down. Frostbite is subtler. Severe frostbite turns fingers and toes black. Your digits actually freeze. When this happens, they need to be amputated to avoid spreading disease to the rest of the body. Mild frostbite can damage nerves permanently, resulting in fingers that can't twiddle about the way they used to. Hats protect the ears; gloves and warm shoes save the fingers and toes. These parts of the body are the most susceptible to frostbite.

The greatest cold myth of them all might be that of President William Henry Harrison, who died after only thirty-one days in office. Legend has it that Harrison didn't wear a hat during his inaugural address on March 4, 1841 and that as a result, he caught a cold that eventually killed him a month later. The legend is somewhat of a double fib. You now know that you don't catch a cold from being in the cold weather. Regardless, Harrison wasn't hampered by any cold he caught on or around his first day on the job. Records reveal that he met with many people during his first month. At night, he would go for long walks and visit local shops. Harrison did seem to have a cold, but, since it was winter, so did many others. He didn't feel sick until March 27, when he developed a fever. He was diagnosed with pneumonia on March 28, and died in the White House on April 4, 1841. Some historians say that Harrison fully recovered from his cold and developed pneumonia afterwards. Colds and pneumonia are two different things caused by two different viruses.

Now, it is indeed true that Harrison rode a horse in the inaugural procession in damp, subfreezing weather and then gave a ninety-minute speech without his hat, gloves, and coat. The long, fiery speech (edited, but perhaps not enough, by Daniel Webster) and the coatless delivery amounted to a political stunt. The sixty-eight-year-old Harrison was out to prove to the American public that he was as fit as a young man. After all, Harrison's political slogan had been "Tippecanoe and Tyler too," a reference to his famous battle with Native Americans in Tippecanoe County, Indiana, in 1811. Harrison's party, the Whigs, painted this rich, Virginian aristocrat as a poor, rugged frontiersman.

What is clear, however, is that exposure to the cold did not cause Harrison's cold or pneumonia. The White House, like any other home, was filled with stale air in the wintertime. Visitors came and went, no doubt full of viruses. Furthermore, Harrison shook a lot of hands before and after the inauguration, hands that could easily have contained cold viruses. Harrison shook so many hands, in fact, that he stopped midway through his inaugural day because his hands were so sore. One other fact rings clear about Harrison: dying after thirty-one days in office probably didn't do much for that rugged image he was striving for.

17

ⴵ ⴵ ⴵ ⴵ ⴵ ⴵ ⴵ

The Ill-Advised War on Bacteria: Are All Bacteria Bad?

Pity the poor bacterium, the Rodney Dangerfield of the uni-
cellular world. It eats our trash, makes soil fertile, turns the
food we swallow into useful vitamins, and yet it gets no respect.
Most people, when you get right down to it, are just plain bigots
when it comes to bacteria. They want to run all two-thousand-plus
species of bacteria out of town just because of a few ornery germs
that can make us sick.

Ridding ourselves of bacteria is a hopeless and foolish endeavor.
Bacteria were likely the first life forms on earth, and they will likely
be the last ones around when the sun starts exploding in a few billion years. Bacteria live in just about every nook and cranny imagi-
nable: in hot springs, on the rims of volcanoes, down underground
sulfur vents, across the frozen continent of Antarctica. Grab a
handful of dirt, from anywhere, and you'll have a handful of bacteria. Bacteria dominate the world. As the late Stephen Jay Gould
wrote, this is not the Age of Man; there was never an Age of the
Dinosaur. We are and always will be living in the Age of Bacteria.

Bacteria are like little, one-celled plants and animals. They are
smaller than most cells in your body. A blood cell is about 5 to 8
microns wide; a thousand microns are in a millimeter. A bacterium
is about 0.5 to 1.5 microns; a sperm is huge at 60 microns. (Viruses
are the smallest, 0.05 microns.) Algae, or blue-green bacteria, have
chlorophyll and need only sunlight and water to survive. All other

bacteria need to eat in the same way animals do. Some can live off inorganic material, such as gases. Most need organic matter, such as dead or living plant and animal tissue. Your body, inside and out, is covered in bacteria, and you should be happy about that. Bacteria outnumber human cells in the body about ten to one.

Human skin contains many species of harmless bacteria. You can take a hot shower, but they aren't going anywhere. They moved in soon after you were born and built up quite a tight-knit community during your childhood. Although these bacteria have no quarrels with you—you are, after all, donating your skin—they are tough on invading bacteria. There's only so much skin to go around, so they are protective of their real estate. Harmful bacteria, what we commonly call germs, will have difficulty gaining a foothold on your skin if your body is already covered with harmless bacteria.

Inside the body, the entire digestive tract is lined with bacteria, from top to, uh, bottom. These bacteria work with the body's own chemicals in breaking down food, converting it to useful vitamins and minerals, and making sure the intestinal walls can absorb nutrients for the blood stream to circulate. Without these bacteria, we could not digest food. Indeed, babies are born relatively bacteria-free, and they are extremely limited in what they can eat. Exposure to bacteria is essential for children to develop a working digestive and immune system. Much like a vaccine—which introduces a dead or weakened virus to your body so that you can build up a resistance to it—bacteria trigger the formation of antibodies. These are proteins in the blood, sort of like foot soldiers, that attack harmful germs that slip past the border of your skin. Without early exposure to bacteria, the body remains ill-prepared.

In fact, some doctors believe that the rising incidence of asthma and allergies in the United States is tied to the relatively sterile world our children live in compared to a generation ago. Children not exposed to bacteria do not receive the germ workout required to make antibodies. More specifically, they do not develop T-helper 1 cells, which make antibodies for allergens. Then along comes a dust or pollen particle. Ongoing research at Tufts School of Medicine and at the Mayo Clinic is showing that, in some cases, asthma and allergies are a reflection of a hyperactive immune system not knowing how to conquer an invading particle.

How did we get so clean? Cleanliness is next to godliness. We've taken this motto and have really run with it. We don't just want our households sparkling. We want them germ-free. According to the Soap and Detergent Association (yes, this is a real association), over three-quarters of liquid soaps and over a quarter of bar soaps on supermarket shelves contain triclosan, an antibiotic that kills most bacteria, both good and bad. The antibacterial craze is also woven into pillowcases and sheets, injected into plastic in children's toys, and even squeezed into toothpaste tubes.

Is any of this necessary? For 99.9 percent of us, no. There is no denying that there are bad bacteria out there. We don't want these germs to invade our bodies. Salmonella (often in eggs), *E. coli* (from meat contaminated with feces), and cholera (in water) can do a number on your intestines and are potentially deadly. Note, however, that antibacterial soaps cannot kill these bacteria; only properly cooking food and treating water can. Colds and flus can knock us out of commission for days and even weeks. Yet colds and flus are caused by viruses, not bacteria, so antibacterial soap won't work here either. Bacteria do cause strep throat, pinkeye, and many types of pneumonia, but regular soap can kill these germs.

So what's the use of antibacterial soap? First, a word on what soap does in general. Soap washes off dirt as well as viruses and bacteria from your body—especially newly contracted germs that haven't had a chance to settle in and multiply. Frequent hand-washing does wonders. If you want to really lower your chances of being infected by harmful bacteria (or the cold virus), wash your hands whenever you think about it and certainly whenever leaving the toilet. This isn't anal retentive, just wise. No one is suggesting you need to scrub fifty times a day until your skin turns raw.

Antibacterial soap lifts off germs, like regular soap. It also leaves a chemical film to kill other bacteria on your skin and prevents bacterial growth for a day or two (no one is sure). Sounds good. The problem is that antibacterial soap doesn't kill 100 percent of a specific group of bacteria. The soap may kill only about 90 percent, leaving behind a strong 10 percent that was able to

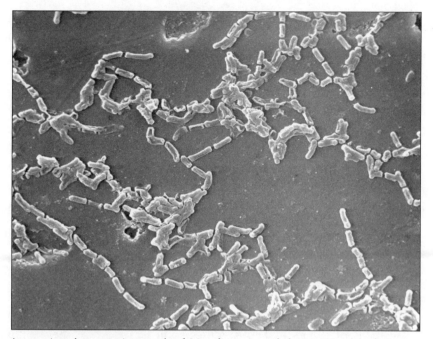

A scanning electron micrograph of *Mycobacterium chelonae*. Friend or foe? Most bacteria are harmless; many are necessary for human life. Antibiotics kill all bacteria and should be used with caution. *Courtesy of CDC/Janice Carr*

resist the triclosan, that antibacterial chemical. Now these bacteria will reproduce, and the next generation will be even more resistant to triclosan. Soon, the triclosan won't work at all because the bacteria have mutated into "superbugs." These bacteria now have an upper hand in wreaking havoc on the body. What's worse, the antibacterial soap has killed the harmless bacteria. This opens up more real estate on the skin for the bad, resistant bacteria to occupy. The same holds true for kitchen countertops. Antibacterial soaps leave a chemical film where resistant bacteria can multiply. Alcohol and bleach kill bacteria and then evaporate; bacteria cannot develop resistance to these chemicals and they are therefore much more practical in killing germs.

Scientists are so worried about the proliferation of antibacterial soaps that they are calling upon the U.S. Congress to outlaw them. You can't blame a person for wanting to be clean, though.

Basic hygiene—washing hands, handling food properly, having access to clean water, quarantining the sick—has increased human life expectancy more so than any medication or surgery technique. In the last hundred years, U.S. life expectancy rose from forty-seven to seventy-two, and much of this can be attributed to the fact that we have cleaned up our act. Up until the mid-1800s, there was little understanding that microscopic bacteria were the cause of so much death. Doctors would perform surgery barehanded. Medical students often moved from working with cadavers to delivering babies without washing their hands. President James Garfield would have survived his assassin's attack had he not developed a bacterial infection from the bare fingers of a doctor probing the wound for the bullet.

Joseph Lister (of Listerine fame) was one of the first to publish research about germ theory and antiseptic techniques, in the 1870s. Lister was largely ignored and even publicly ridiculed until the turn of the century. Cleanliness finally clicked with the public health movement of the early 1900s. Cities soon learned that they could prevent cholera outbreaks by providing fresh drinking and cooking water, for the disease is caused by bacteria in raw sewage, which often filled the back alleys of large cities such as New York and Chicago. Garbage collection and removal was a surefire method of ridding communities of diphtheria and scarlet fever, both of which were caused by bacteria breeding on trash. Better sewage and garbage collection also got rid of flies, whose suction-cup legs carried bacteria from feces and trash onto tabletops and food.

Today, Americans will apparently settle for feces in their food supply, but they demand antibacterial soaps. Most bacterial infections in the United States are food borne: salmonella, listeria, and *E. coli*. The bacteria proliferate in factories where food is mass-produced. The majority of meat produced in this way contains fecal material, a trend that hasn't existed for nearly one hundred years. We cannot wash food in triclosan. The best we can hope for is better food-safety practices and an emphasis on smaller, local meat producers instead of centrally located conglomerates that process tons of meat each hour. Many health experts suggest irradiating food, dousing it with low-level radiation to kill bacteria.

Knowing how much Americans fear radiation, though (see chapter 18), they use the euphemism "cold pasteurization." This process is fine at the slaughterhouse, but the food can be contaminated during the long, long trip from the slaughterhouse to your kitchen. Perhaps home irradiation kits?

The greatest abuse of antibiotics occurs in the livestock industry. Cattle, hogs, and chicken are pumped up with antibiotics to prevent disease that would otherwise be rampant in the cramped, stressed confines in which these animals are raised. Approximately 80 percent of all antibiotics are used for raising livestock, and this poses the greatest threat for bacterial resistance. Smaller farms and less crowding would eliminate the need for animal antibiotics. Another great abuse concerns doctors who prescribe antibiotics haphazardly to their patients, who most of the time don't need them. Colds and flus are viral infections, and antibiotics are useless in curing them. Antibiotics often are prescribed to placate nervous patients. The anthrax scare of late 2001 prompted many Americans to stockpile antibiotics, namely a product named Cipro, because this antibiotic *is* effective at fighting anthrax infection. Nearly all of the estimated tens of thousands of people who took Cipro "just in case" never had anthrax. Thousands more hoarded Cipro, and the fear is they will self-administer the drug at the onset of a cold.

In China, where antibiotics are readily available without instructions for use and abuse is widespread, the majority of bacteria causing urinary tract infections and other life-threatening diseases is resistant to fluoroquinolone, the family of antibiotics that includes Cipro. The same will likely happen in the United States, experts concede. Without Cipro and other antibiotics working to kill harmful bacteria, we will be as helpless as we were a hundred years ago in fighting disease. Already, tuberculosis and other infections once under control are now largely resistant to most antibiotics. Stuart Levy of Tufts Medical School, president of the Alliance for the Prudent Use of Antibiotics, documents the abuse in his 2002 book, *The Antibiotic Paradox*. Antibiotics are powerful, toxic drugs. You cannot simply pop these pills like vitamins to bolster your health in a time of increased risk, regardless of how much this news bugs you.

18

ᵿ ᵿ ᵿ ᵿ ᵿ ᵿ

Radiating Misperception: Radiation, Pro and Con

Have you ever heard of a medical procedure called nuclear magnetic resonance? Maybe not. How about magnetic resonance imaging, or MRI? Yes? MRIs are quite useful for taking photographs of soft tissue such as the brain and other organs to look for tumors and abnormalities. Essentially, these machines use magnets and pulses of low-energy radio-wave radiation to create images from energized hydrogen atoms in our body's water and fat molecules. You no doubt have seen these futuristic-looking devices—big white things in which a patient lies on his or her back and slides into the tunnel that cuts through the heart of the machine.

Magnetic resonance imaging was originally called nuclear magnetic resonance. Marketing research quickly revealed that the public was scared of the word *nuclear* and was hesitant to undergo the new life-saving procedure because of that. Faced with trashing a potential billion-dollar industry over the connotation of one word, the nuclear magnetic resonance industry promptly dropped the "nuclear" part. The word *magnetic* was fine; people have magnets on their refrigerator. But *nuclear* means "radiation," and *radiation*, to so many of us, means "cancer" and "death."

This is no laughing matter. We are petrified of radiation, not understanding that most forms of radiation are safe. Radar weather towers that would have provided early hurricane warnings have been either decommissioned or never built because local residents feared the radiation from the towers more than they feared the real threat of hundred-mile-an-hour winds carrying glass shrapnel from storm-struck storefront windows. This radiation would be several orders of magnitude lower than the radiation dose they get from the sun each day. The cell phone industry is also getting walloped these days over the radiation issue. Cell phone radiation cannot be healthy, people are saying. Must be causing brain tumors, they say.

Back in the 1960s and 1970s, the worry was microwave ovens. The microwave oven industry started off at a snail's pace; its only customers were restaurants. The general public didn't want to cook their own food with radiation from microwaves, not understanding that humans have cooked food with another form of radiation since the dawn of the campfire. Radiation is, after all, energy—energy that travels in waves (such as the infrared energy from a stovetop) or as subatomic particles.

The electromagnetic spectrum is pure radiation, comprising radio waves, microwaves, infrared light, visible light, ultraviolet radiation, X rays, and gamma rays. Some sections of the spectrum must have a good PR agent. For instance, no one thinks radio waves are harmful, except maybe the kind that carry the tunes from those vapid top-40 pop music stations. Deadly stuff, if you listen long enough. The infrared is indisputably cool, allowing snipers and spies to see at nighttime with infrared goggles because all objects that have heat (humans, buildings) emit infrared radiation whether the light switch is on or off. Visible light is home to the rainbow. You can't knock visible radiation. Energetic radiation—UV, X ray and those crazy Greek letters gamma, alpha, and beta—well, they can be trouble. More on this type of radiation later.

Microwaves are a form of low-energy radiation. They cook meals when concentrated in an oven by vibrating the water molecules in food, which creates heat. This is an efficient method of cooking because heat (and thus cooking) is localized within the food. On a stovetop, a gas flame or electricity produces infrared radiation, which transfers energy (heat) to a frying pan, which in turn transfers heat to the outside of the food. The end result is the same: radiation produces heat, and heat breaks chemical bonds in food in a process we commonly call cooking. Microwaves just do it faster.

Paul Brodeur, an investigative reporter with the *New Yorker* magazine, helped fuel the microwave scare in the 1970s. In his articles and a subsequent book with the admittedly clever title of *The Zapping of America,* he relayed amazing statistics such as the fact that electromagnetic radiation from microwaves, radar, and television has increased 100 million times since World War II. Sounds scary and it's likely true, but this amount is still minuscule compared to the natural background radiation emitted from the sun and even our own bodies. Countless medical studies have since proved that microwave ovens don't cause cancer. Most folks today comfortably use the microwave oven, their fears distilled, perhaps, in the comic description of the process as "nuking" one's food. No one is getting sick from microwaves.

The next scare, circa 1979, was from electric power lines, transmitters of electromagnetic force, or EMF radiation. Several children in a Denver, Colorado, neighborhood had developed leukemia. One epidemiologist visited the area to search for possible environmental contaminants and noticed that the children's homes were clustered around power lines. Could the power lines have caused leukemia, a blood cancer? Don't know; it was certainly worth investigating. And so they investigated . . . for the next eighteen years. Nothing. But, we were talking about sick kids and big, bad power companies. This made for good television news. Paul Brodeur, the *New Yorker*'s microwave man, took up this crusade with magazine articles and another book, *Currents of Death,* a follow-up to his antimicrowave tour de force.

A connection between low-level radiation from power lines (which are even less energetic than microwaves) and leukemia was tenuous at best. There was no known biological mechanism for this type of radiation to cause damage to DNA, the root cause of cancer. Also, millions of people lived near other power lines, and the children there didn't have any more leukemia than kids elsewhere. Activists, nonetheless, charged that the power companies and the U.S. Department of Energy were involved in a masterful cover-up. Power lines weren't the only source of deadly radiation, they said. Everything electric emitted radiation—electric blankets, televisions, phones, lighting—and all were suspect. The power industry was out to protect its fortunes, the argument went, and naturally denied any adverse health effects from the transmission and use of electricity. It was not clear whether opponents of EMF wanted us to forgo electricity and resort to the oil lamp (once fueled by whale oil, which is why the whales are nearly extinct.)

The public bought into the fear . . . or at least Hollywood did. In Eddie Murphy's 1992 flick, *The Distinguished Gentleman*, a con man elected to Congress finds salvation as an environmentalist battling a power company, whose power lines near a playground cause a little boy's cancer. Both the National Academy of Sciences and the National Institutes of Health decided to settle the power-line radiation issue once and for all. In 1996, the Academy—a stuffy albeit terribly brainy Who's Who of modern science—concluded after an exhaustive three-year review that there was no connection between power lines and any type of cancer. In 1997, NIH chimed in with the results of a seven-year, comprehensive (read "expensive") study, also finding no connection. In 1999, the Canadians nailed the lid on the coffin with their nationwide study finding the same nonassociation. According to the White House Science Office, the total cost of the power-line radiation scare was over $25 billion. For that money, we could have sent humans to Mars, or, perhaps more practically in this case, found a cure for leukemia. Robert Park provides a nice overview of this in his 2000 book, *Voodoo Science*.

What is it about radiation that worries people? Many folks seem to equate all types of radiation with the dangerous kind, called ionizing radiation. This type of radiation is energetic enough to knock an electron loose from an atom. Many types of radiation pass through our bodies all the time. Although you cannot shine a flashlight (visible light) through your chest, radio waves and microwaves will pass through easily enough. Ionizing radiation also passes through you, but it can damage the atoms in your cells as it travels, knocking loose an electron from a DNA molecule. Ultraviolet light is ionizing radiation; too much of it causes skin cancer. X rays and gamma rays are also ionizing. Too many X-ray exams can lead to organ cancers. Fortunately, the nastiest forms of ionizing radiation are produced in deep space, and the earth's atmosphere blocks most of this from reaching the earth's surface (although that hole in the ozone layer is letting in more UV).

Radio waves, microwaves, and infrared and visible light, no matter how plentiful, cannot knock any electrons loose to cause cellular damage. This is a key property of quantum physics. Only a photon (light particle) of a certain energy can bump an electron loose, and that kind of energy doesn't start until the upper end of the UV spectrum. Think of photons as baseballs and an electron as a window in a house across the street. Radio waves do not have enough energy to make it across the street. You can throw a million of them; nothing's going to shatter that window. UV, X-ray, and gamma-ray photons have enough energy to make it across the street and send the old man behind the window running after you.

Over 80 percent of the ionizing radiation we encounter day to day comes from natural sources: cosmic rays, which are atomic particles from space; and alpha and beta particles from radioactive gas, namely radon. Ionizing radiation is actually hard to avoid. Radon gas, for example, accounts for nearly 70 percent of natural ionizing radiation. This stuff comes from the decay of uranium in soil and percolates up into the open air or into basements through cracks in the floor. Radon becomes a health hazard when it gets trapped in a building and accumulates. We also get a small dose of radiation from cosmic rays on international flights, when the jet plane reaches altitudes of around twenty-five thousand feet and higher.

Medical X rays account for about all the rest of our ionizing radiation exposure. We worry about ionizing radiation, but nearly 80 percent of it is unavoidable. Clearly we don't want any extra ionizing radiation. Uranium miners suffered through all sorts of cancers from working with radioactive uranium without protection. Early on, the mining industry did not compensate them or their families for their deteriorated lives and deaths. Also, the United States exploded several nuclear bombs in the South Pacific after World War II that caused untold sickness and death for Pacific Islanders. Radon gas causes thousands of lung cancers per year in America, a low but not insignificant number. Aside from

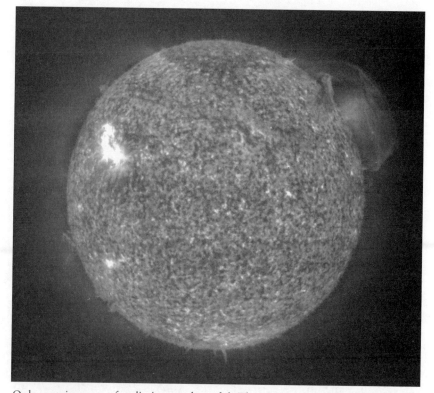

Only certain types of radiation are harmful. The sun's radiation, called light, makes life possible; and it is millions of times more abundant than the cell phone radiation that some folks are worried about these days. *Courtesy of NASA/Solar & Heliospheric Observatory*

these instances, most folks usually do not have to worry about yearly exposure to ionizing radiation.

The ionizing radiation from nuclear energy, however, is cause for alarm. The problem with nuclear energy is that the spent fuel, the "ash," is radioactive and there is no place to store it. Some groups say nuclear energy is clean by virtue of an absence of smokestacks. Coal burning would be clean, too, by that standard, if we collected the smoke from the smokestack and stored it in barrels. This is what happens at a nuclear power plant; the ash is collected. By conservative estimates, the spent fuel in barrels is radioactive and potentially deadly for at least 500 years. Some push that number up to 10,000 years. Regardless, there is no place to store anything safely for 500 years. Empires fall. Look at Rome; look at the Soviet Union. The United States plans to place all of its spent nuclear fuel deep in Nevada's Yucca Mountain. When the United States empire collapses, who will monitor the waste? What would have been the impact on American health had the Iroquois nation filled the Appalachian mountains with poison 500 years ago? Few of us are being poisoned by nuclear radiation today. And barring an accident, nuclear energy is safer than coal, the mining and burning of which causes tens of thousands of deaths each year. The potential for nuclear danger is great, though, so the fear of nuclear energy is far from irrational.

Not so for cell phone radiation. This *is* irrational. The radiation here is nonionizing radio waves. No one bats an eye when teenagers or joggers wear radio headphones. This is the same radiation as that collected and transmitted by cell phones, only at a slightly different frequency. The cell phone scare likely won't go away for several more years, as more and more people use cell phones and it becomes obvious that no one is getting sick. This is how the fear of microwave ovens was put to rest. Two large studies published nearly simultaneously in *The New England Journal of Medicine* and in the *Journal of the American Medical Society* in December 2000 found no increased risk of brain tumors. A much larger European study with more subjects and a longer timeline, to be released at the end of 2002, is suggesting the same thing. A huge study was planned in the United States that would have compiled data on the cell phone use by millions of Americans (by looking at customer records) and compared it to brain tumor cases. This would have sealed the case for cell phone safety, but one person sued, saying the study was an invasion of privacy. Only in America.

19

🜍 🜍 🜍 🜍 🜍 🜍 🜍

Swimming with Sharks:
Sharks and Cancer

Even sharks that never smoke cigarettes can get cancer. Why am I telling you this? Some think that a shark's natural abundance of cartilage keeps the creatures immune to cancer. A multimillion-dollar shark cartilage industry has grown up around this myth; they market the stuff in pill form as an alternative cancer cure. It turns out, in a tragic bout of irony, that not only do sharks get cancer, they can also get cartilage cancer.

Just when you thought it was safe to go back into the water . . . (strike up the scary music) . . . here come the shark cartilage hunters. At least that's the horror movie that sharks live with every day. Ever since the 1992 book, *Sharks Don't Get Cancer,* by William Lane—and the report by the television news program *60 Minutes* that trumpeted it in 1993—sharks have had one more thing to worry about. True, they've been loathed as man-eaters since the days of the movie *Jaws*. For centuries before this, certain Asian fisherman have killed sharks for the fins to make soup, dumping the carcasses back into the ocean. Now, with many species of shark either endangered or darn close to it, health-food gurus are netting them for their cartilage. Strange. The same people who object to rhino horns being ground into an aphrodisiac, or the last patches of rain forest being ripped up by multinational corporations, have no problem with bottled shark cartilage. It's right up there on the shelf with vitamin C.

Cartilage is the soft tissue that cushions the joints in birds and mammals, the regions where bones meet. Sharks and their cousins, rays and skates, are different from fish and other animals in that they have no bones but rather a cartilage "skeleton." Cartilage does have chemicals that halt cancer growth, as we shall see. And pound for pound, sharks have more cartilage than farm animals such as cows or chickens. So sharks are the target of the cartilage harvest. There is no evidence, however, that ingesting cartilage in pill form—where it lands in a puddle of stomach acid and proceeds along its fantastic journey through the gastrointestinal tract—can travel to the site of cancer damage and work its magic. The active ingredients in shark cartilage are too large to be absorbed in the bloodstream. They pass on through. The Federal Trade Commission, in fact, has sued shark cartilage packagers such as Lane Labs-USA, for making unsupported claims about the anticancer properties of shark cartilage. (Yep, the guy who wrote the book sells the cartilage. That's standard practice in the world of alternative medicine.)

Cartilage might curb cancer if it can get to the right spot in the right concentration at the right time. The National Institutes of Health is sponsoring a large health study on shark cartilage, which is partially funded by the cartilage industry as a result of the FTC lawsuit. The vast majority of past studies have shown that shark cartilage doesn't cure cancer. Neither the National Cancer Institute nor the American Cancer Society recommends shark cartilage. Yes, there have been testimonials, usually unfounded. Yes, cancer patients once thought to be terminal have recovered fully after taking shark cartilage. But a lot more miraculous things have happened in this world. Cancer remission can occur spontaneously. If you are playing the harmonica when remission hits, then harmonica playing cures cancer. Such is the logic of every miracle cancer cure.

How did it all start? Doctors first noticed decades ago that cartilage from cows had properties that halted cancer-cell reproduction. The cartilage prevented angiogenesis, the growth of new blood vessels. Normally this would be a bad thing, for the body needs new blood vessels when repairing wounds or, in females, nurturing a fetus. But cancer cells also need new blood vessels to fuel their growth. The anti-angiogenesis properties of cartilage starve cancer cells of water, nutrients, and oxygen. Once again, enter the book *Sharks Don't Get Cancer*. The author does admit in his book that sharks do get cancer, just not that often. An alternative title, *Sharks Get Cancer Sometimes But Read This Book Anyway*, clearly doesn't have the same appeal. The science community, though, has really given William Lane a hard time about the book, its title, and his stance.

Lane's premise is that sharks, with their cartilage skeletons, are loaded with these anticancer properties. This is why they have a lower rate of cancer than humans. There are two main problems with this logic. First, the same anti-angiogenesis in the cartilage that may stop cancer would also stop wound healing and all the "good" blood-vessel building. Second, sharks do get cancer. They even get cancer of the cartilage. Sharks may get cancer more often than we think. It's hard to tell. Sharks are notorious for not visiting their doctors for regular cancer screening. No one knows the true rate of cancer among sharks. In interviews, Lane has placed the rate at "about one in a million," perhaps a figurative statistic. In his book, Lane documents 30 tumors in 7,500 sharks catalogued by the Smithsonian, an organization that saves dead things. Thirty out of 7,500 is about one in 250. Admittedly this is better than a human's *lifetime* risk of getting cancer, which is about one in four, but this isn't a fair comparison. Human cancer rates vary greatly with age, lifestyle, environment, and socioeconomic status. The U.S. 1998 incidence rate for all cancers was about 400 per 100,000 people, or one in 25. Perhaps more sharks would get cancer if they didn't die young from starvation or wholesale slaughter.

Even if sharks get cancer less often than the rest of us, that doesn't imply that the known anticancer property of cartilage is the reason. Sharks don't have bones, which makes them unique. Bones are the source of marrow, which produces blood cells and other types of disease-fighting cells for the immune system. These cells mature in the bones and then are released into the bloodstream. That takes a little time. In sharks, disease-fighting cells are produced in the spleen, thymus, and tissues associated with the gonads and esophagus. Research is revealing that the disease-fighting cells actually mature in the bloodstream. That is, the soldiers are already in the field and not in the bone-barracks when disease or sickness strikes. Maybe this keeps sharks healthier than other animals. We just don't know.

The book *Sharks Don't Get Cancer,* also makes hay out of the fact that sharks have survived unchanged for the past 400 million years and that they are the "ultimate biological machines" that "do not need to sleep or rest." Well, sharks do sleep. Just as with the cancer myth, research has revealed that sharks sleep in unique ways. Just because we don't see something at first doesn't mean that it doesn't exist. As for the longevity argument, cockroaches have been around for a long time too. No cartilage but plenty of protein; I recommend them pan-fried in sesame oil and red pepper.

The scientific attack against Lane's claims continues, perhaps spurred by the publication of his cocky sequel, *Sharks Still Don't Get Cancer.* This time *60 Minutes* backed off from promoting Lane. In 1993, they followed him to Cuba, where allegedly terminal cancer patients who received shark-cartilage preparations felt better after several weeks of treatment. This Cuban study, along with a Mexican study, is cited widely in *Sharks Don't Get Cancer.* Of course, "feeling better" is not synonymous with "cured of cancer." The National Cancer Institute subsequently reviewed the Cuban data and found it to be "incomplete and unimpressive." (Although solid health studies are conducted in developing countries, you have to question why the cartilage study was not performed in a more established setting. Why Cuba?)

Cancer specialists are not so concerned with the fate of sharks or the money being made from their cartilage. Their main worry is that cancer spreads so quickly. When detected early, cancer can be treated or surgically removed. When a patient elects to forgo surgery, medication, or radiation treatment—for whatever sound reason—and instead chooses to take shark cartilage, that patient runs the risk of dying. Not all cancer treatments are the same: there are more effective remedies, and there are less effective remedies. And then there are shams, such as the "apricot pit" cancer cure of the 1970s, laetrile, which only imparted false hopes and drained pocketbooks for the patients who rushed to Tijuana for that treatment.

Shark cartilage as a cancer cure? Beware of the sharks who tell you this is true.
Courtesy of National Oceanic & Atmospheric Administration

The FTC action has made it illegal for shark-cartilage dietary supplements to mention anything about cancer on the label—at least in the United States. Often, shark cartilage is marketed as a general elixir. The word on the street, naturally, is that shark cartilage attacks cancer. That's what the title *Sharks Don't Get Cancer* accomplished, and that's the chime of word-of-mouth advertising. Maybe cartilage holds great promise in the treatment of cancer. More research is definitely warranted. We all can't wait to hear about the NIH results. Any treatment is better than god-awful chemotherapy—if it works!

This may be the logical place to add that another animal, the polar bear, doesn't seem to develop colon cancer. The bears' high-fat, low-fiber diet would surely doom a human being, yet they stay cancer free. A naturally occurring acid called UDCA, which exists in high concentration in the bile of polar bears but only in low concentrations in human, might prevent colon cancer. Scientists are investigating this, fortunately without killing polar bears. And just how does one go about screening polar bears for colon cancer? Very carefully.

20

𝌆 𝌆 𝌆 𝌆 𝌆 𝌆

Mutating Misconceptions:
What Your Genes Say
about Your Future Health

Nervous? Just because your mother or father died of liver cancer doesn't mean you will. Rarely do folks inherit genes that guarantee a disease. At best, your genes may leave you slightly predisposed to a given disease, which means you might be more likely than others to get that disease. This, in turn, means that when a cancer-causing chemical or an influx of fatty foods enters into your body, you may have a harder time than others in fighting the harmful effects of these substances. But you are not doomed. Genes are the scapegoats of the twenty-first century.

In the United States the top 10 killers are, in order: cardiovascular disease, cancer, stroke, bronchitis-emphysema-asthma, accidents, flu-pneumonia, diabetes, suicide, kidney disease, and liver disease. The risks of any disease, however, vary with age, race, and gender. Black male youths in large cities are much more likely to die from gun violence than from stroke. Cervical cancer is five times more common among Vietnamese immigrants than among white Americans.

All of the top killers are preventable, or their risk can be greatly reduced, especially violence among young people. Only the rarest of cancers are purely genetic. For example, the average American has about a 5 percent chance of developing colon cancer over his or her lifetime, according to the National Cancer Institute. Colon cancer is nearly a sure thing, though, for individuals with a hereditary disease called familial adenomatous polyposis (FAP). People with FAP develop hundreds and even thousands of potentially cancerous polyps in their colon and rectum. Polyps in general have about a 1-in-50 chance of turning cancerous, so the more polyps you have, the worse things get. About one in every million humans have FAP, as dangerous and often tragic as it is, and FAP accounts for fewer than 0.1 percent of all colon cancers. Exercise, diet, and all that good stuff cannot stop the polyps from developing for people with FAP. Fortunately, modern-day colonoscopy techniques can find and remove polyps before they turn cancerous; and a drug called celecoxib can help prevent polyps from growing and turning cancerous by regulating a gene called COX-2.

A healthy lifestyle can indeed greatly minimize the risk of colon cancer for the 99-plus percent of the world population without FAP. A low-meat diet with plenty of water (a half gallon a day) knocks your 5 percent risk down to about 1 percent. Colon examinations every five years or so after age forty minimize the risk even more, because colon cancer—unlike most other cancers—takes a long time to develop from existing polyps. Colon cancer, like skin cancer, is one of the most preventable of all cancers. Many health experts say that no one should be dying from colon cancer, yet it is the second leading cancer killer in the United States, behind lung cancer. Charles Schulz, creator of *Peanuts,* died of colon cancer. If your parents or siblings died of colon cancer, you are not destined to follow. You are considered "high risk," which really means "at a higher risk than others." So you need only be more diligent about colon exams, low-meat diets, and the like. In reality, the majority of colon cancer cases are those individuals with little or no family history of colon disease. The disease takes them by surprise.

We can go down the list of the top-ten killers and rule out genetics one by one. Lots of folks' fathers have died of heart attacks and clogged arteries. These older men might have also drunk bacon grease every day for breakfast. That is, there was less worry years ago about how diet affected longevity, for most men were dying by age sixty-five. Many of us, including myself, are genetically predisposed to a risk of high cholesterol levels and a buildup of fatty droplets in the arteries, a precursor to cardiovascular disease. Many of us, including myself, had a grandfather who died at age forty-nine and a father who died at age sixty-two of heart disease. As with colon-cancer prevention, though, genes do not determine our fate. The evidence is overwhelming that low-fat, vegetable-dominated (but not necessarily vegetarian) diets coupled with even casual exercise dramatically lower all of the risk factors associated with cardiovascular disease, such as good/bad cholesterol ratios and high blood pressure, regardless of genetic predisposition. Having the "bad heart" gene only means "be more careful than others."

With all the common cancers—those attacking the lung, prostate, breast, and organs as well as the colon—genetics plays far less of a role, if any, than environmental factors such as cigarette smoking, high-fat and salty diets, inactivity, occupational hazards, and access to health care. Stroke is akin to cardiovascular disease in the brain, so the same prevention methods apply. Bronchitis and emphysema deaths are largely related to cigarette smoking, not genetics. Asthma seems to be genetic, although more children seem to have this condition today than ever before. No one is sure why. Experts point to air pollution, but the air is really the cleanest it has been in cities for about three hundred years and—when considering dust, fleas, rodents, animal waste, chamber pots, and smoke and particulate from fireplaces—cleaner in the home than at any time in history. Nevertheless, aerobic exercise can greatly strengthen the lungs and minimize the risk of asthma death regardless of genetics.

Accidents, such as falling down the steps or running a red light and smashing into another car, are clearly preventable unless you inherit the "two left feet" or "jerk" gene, respectively. No one is genetically predisposed to a higher risk of catching a flu or developing pneumonia. These diseases are caused by viruses and bacteria. The elderly and those with weakened immune systems (such as HIV patients and cancer patients on chemotherapy) are the groups at highest risk of flu and pneumonia death.

Type II diabetes rates have risen by a third since 1990 in the United States and are climbing in other industrialized countries as well. Roughly 16 million Americans have the disease, and another 10 million are at risk. Type I diabetes, often called juvenile diabetes, is partially genetic and accounts for about 5 percent of all diabetes cases. Type II, also called adult-onset diabetes, is almost entirely associated with diet and obesity, not genetics. Normally, only adults get this disease, usually after age forty. However, children, who are growing heavier and heavier statistically each year, are now developing adult-onset diabetes before they reach their teenage years. The National Institute of Diabetes and Digestive and Kidney Diseases announced in August 2001 that modest lifestyle changes—eating less fat, walking thirty minutes a day, and losing a little weight—can cut the incidence of diabetes by more than half for people who are most at risk. The results from these changes were better than results produced by metformin, the diabetes-prevention drug. Harvard researchers published a report in *The New England Journal of Medicine* in September 2001 stating that 91 percent of Type II diabetes cases could be attributed to lifestyle: smoking, obesity, lack of exercise, and poor diet.

Suicide, surprisingly common and always tragic, is loosely associated with genetics, but one can attain a healthy mental outlook through positive social factors, such as companionship, career satisfaction, community involvement, religious affiliation, and avoidance of drugs and alcohol. Kidney and liver diseases, also, are loosely associated with genetics but are far more likely to be caused by environmental factors, namely exposure to toxins from work or in one's diet. Alcohol abuse is the leading factor in liver disease. Blood filters through the liver, which detoxifies harmful chemicals, and through the kidneys, which remove certain chemicals for excretion in urine. An influx of toxins (industrial solvents, poisonous herbs or food, metals such as mercury) can overwhelm these organs, compromise their ability to function, and lead to their demise.

Among other feared diseases, Alzheimer's is only partially genetic. About 5 to 10 percent of Alzheimer's patients have the genetic form of the disease, and this usually develops between ages thirty and fifty. Amyotrophic lateral sclerosis (ALS or Lou Gehrig's disease) affects about 1 in 100,000 people, and about 5 to 10 percent of the cases are due to genetic mutations. Otherwise, the disease seems to strike randomly and is associated with an overabundance of glutamate, a neurotransmitter. Parkinson's disease, affecting 1 in 500 people, has no known genetic link and, actually, no known cause. Multiple sclerosis (MS) affects about 1 in 1,600 people and does not greatly shorten life span. The disease is more prevalent farther from the equator, in the northernmost and southernmost latitudes. Here, genes do play a significant role, but no one is sure how much of a role. MS might also be caused by a viruslike organism, which would jibe with the geographical distribution.

Rarer diseases often have an obvious genetic connection. Huntington's disease, which took Woody Guthrie, is inherited as a single faulty gene on chromosome #4. This tragic brain-wasting disease affects about one in a million people. However tragic these progressive diseases are, the risks of developing the diseases are small, and the therapies are improving each day.

Science's search for the genes responsible for certain diseases is noble, yet often tinged with folly. Finding the obesity gene, for example, will only lead to therapies that will allow us to overeat and never exercise, as we sit and sit, knowing that a pill can erase years of abuse. We know that diet and exercise can reduce the risk of high cholesterol and lower levels once high, yet we seem to rely instead on magic bullets to do the job, such as statin drugs, which can cause liver damage. We know that diet and exercise can lower the risk of diabetes better than metformin, which also has side effects. Our reliance on science to solve problems we can easily manage with caveman-era technology is among the reasons many well-educated individuals venture into the world of Ayurveda or aromatherapy for alternative cures.

Nonetheless, we can only benefit from the knowledge of why some people—stripped down to the genome level—seem more resistant to disease than others are. Wouldn't it be nice if we could all live to a healthy old age through a combination of practical and responsible prevention *and* genetic therapy?

PART IV

᳅ ᳅ ᳅ ᳅ ᳅ ᳅ ᳅ ᳅ ᳅ ᳅ ᳅

Eating It Up

What is food to one is to others bitter poison.
—Titus Lucretius Carus (93–55 B.C.E.)

Many of the myths about nutrition come from the notion that
nature cares about us. Nature doesn't care about us at all. She'll
wipe us out in a second if we lean too far over a canyon's edge or
stand too close to the waves breaking along the shore. Nature's so-
called foods grow and reproduce with little concern about human
health and welfare. All an apple tree cares about is passing on its
genetic information to another generation. Nature doesn't consider
apples as food. If humans can eat the apples to survive, fine. If
only other animals can digest apples, then that's fine, too. Most of
what grows in the wild is inedible to humans, and some of it—
such as half of all mushroom species—is deadly. Some plant foods
are edible raw; others need to be cooked. Sometimes only certain
parts of an edible plant can be eaten and other parts are poisonous. There's no consistency to nature's way. Nature has no intent,
so there is no such concept as "the way nature intended it."
Humans must take what they can get. If we ate only the food that
nature has prepared for us "as is," as other animals do, we'd be
dead.

113

21

ᣮ ᣮ ᣮ ᣮ ᣮ ᣮ

Learning Your Alpha-Beta-Carotenes: Antioxidants, Pro and Con

It used to be so simple, a battle between good and evil. Rogue chemicals called free radicals roam about the body like brazen street punks, smashing cellular walls and roughing up innocent DNA molecules, causing cancers and the diseases of middle and old age. Their flagrant disregard for the law would continue unchecked if it weren't for swashbuckling antioxidants swooping in on the wings of vitamin supplements, disarming the free radicals of their menacing electrons and converting them into respectable molecular citizens.

At least that's how the theory went. But the body, it seems, is not governed by a Hollywood "B" script. Free radicals are as good as they are bad, and too many antioxidants may harm the body. You know many antioxidants by name: vitamins C and E; beta-carotene, a form of vitamin A; and selenium, to list a few. We are being force-fed them in high doses as if they were a proven magic bullet. They're not. It's a very complicated story.

It is true that antioxidants serve as sort of a rust protector for the body, stopping a process called oxidation. Important molecules in the body, such as the ones that form artery walls, become oxidized when they lose an electron. Once oxidized, they become unstable and easily break apart. The culprit, without a doubt, is the

115

free radical. Free radicals are highly reactive molecules or single atoms with unpaired electrons looking for a mate. So they steal an electron from the first thing they encounter, perhaps a cell wall or DNA. As free-radical damage mounts, cells can no longer perform properly. Disease sets in. An excess of free radicals has been cited in cardiovascular disease, Alzheimer's disease, Parkinson's disease, and cancer. Aging itself has been defined as a gradual accumulation of free-radical damage.

Yet, free radicals are necessary for life. The body's ability to turn air and food into chemical energy depends on a chain reaction of free radicals. They are also a crucial part of the immune system, floating through the veins and attacking foreign invaders. Hydrogen peroxide is a prime example of a free radical. Your blood actually contains trace amounts of hydrogen peroxide, an internal germ fighter. In fact, you could not fight bacteria without free radicals.

Free-radical production is a natural byproduct of breathing. You cannot avoid it. The mitochondria—the cell's power plants—use oxygen gas, O_2, to generate energy. That O_2 converts to carbon dioxide (CO_2) in the process; but sometimes a cousin of O_2 forms, called a superoxide radical, which is like O_2 only missing an electron. Superoxide is one of the most common free radicals, along with hydrogen peroxide. Antioxidants, through a multistep process, react with free radicals and convert them to benign molecules such as water and oxygen gas. A balancing act emerges. The body hopes to avoid excessive free-radical production, but it certainly doesn't want to mop up all the free radicals. A diet rich in fruits, vegetables, nuts, and some meats supplies most people with the antioxidants needed to walk this tightrope. Most doctors agree that few people in modern American society need to boost their intake of the common antioxidants, such as vitamin C and beta-carotene. The value of vitamin E is still up in the air, but it doesn't look so good, either; more on that later.

Nevertheless, up to 30 percent of the population is taking some form of these vitamins as supplements, according to the American Heart Association. Antioxidants are a billion-dollar business; Americans were spending over $30 billion on dietary supplements by the close of the century, and nearly $2 billion of that amount was for vitamins E and C, beta-carotene, and selenium, according to *Nutrition Business Journal.* What do we care that none of the antioxidant health claims has been proven true? Protect against cancer, slow aging, prevent heart attacks? For every study that shows benefits, there is another study that doesn't.

Having too few antioxidants seems to be a bad thing, studies indicate. One health study, published in 1983 in the British medical journal *The Lancet,* found that people with low selenium levels were twice as likely to develop cancer compared to people with normal levels. Another study, published in 1986 in *The New England Journal of Medicine (NEJM),* found that patients with a certain type of lung cancer were four times more likely to be deficient in beta-carotene compared to a control group. A 1989 study from the Netherlands associated low selenium levels with an increased risk of heart attacks. More convincingly, the Harvard-based Physicians Health Study—which has recorded the lifestyles of some 50,000 male health professionals for the past 15 years—found that men who ate a diet rich in vitamin E (from nuts, seeds, and soybeans) were half as likely to develop heart disease compared to those with very low levels of dietary vitamin E. The benefit of boosted levels of antioxidants, however, beyond what diet can supply, has been a bit harder to demonstrate.

Taking generous doses of antioxidants showed some promise early on. Skin-cancer patients given daily selenium supplements were twice as likely not to die from cancer as those patients not given selenium, as reported in the *Journal of the American Medical Association* in 1996. This was a multicenter, double-blind, randomized, placebo-controlled study with over 1,300 patients—all the markings of good science. The findings were so dramatic, wrote the authors, that they stopped the study after six years so that all patients could benefit from the selenium supplement. Other studies showed similar positive results: Vitamin E lowered risk of prostate cancer, postponed the onset of debilitating Alzheimer's symptoms, delayed cataracts, and slowed the progress of coronary artery disease. Vitamin C could, indirectly, stave off blindness, kidney failure, and amputation among diabetics. Extra selenium, a mineral needed only in trace quantities, reduced the risk of prostate, colorectal, and lung cancer.

Concurrently, over the years came the neutral and even negative reports about antioxidant supplements. One study, reported in *The New England Journal of Medicine* in 1994, found that male Finnish smokers were 18 percent more likely to develop lung cancer after taking a beta-carotene supplement. In 1997 *The Lancet* published a study of nearly 2,000 men receiving vitamin E or beta-carotene after suffering their first heart attack. There were significantly more deaths from heart disease in the beta-carotene group and a trend toward more deaths in the vitamin E group compared to the placebo group, according to the study report. Other studies showed similar negative results: no evidence that vitamins C and E or beta-carotene prevented colorectal cancer; no evidence that these "big three" prevented arteries from reclogging after angioplasty; no evidence that beta-carotene prevented cancer or heart disease in over 22,000 physicians over 12 years; no evidence that extra selenium prevented cancer in 60,000 nurses; and more bad news for smokers taking beta-carotene, this time with a 28 percent higher incidence of lung cancer. These studies were reported in *The New England Journal of Medicine* from 1994 to 1997.

Criticisms naturally flowed back and forth. The big Finnish study showing adverse effects from beta-carotene could not rule out the fact that these men might already have had cancer in its earliest stages, said the proantioxidant crowd. Those big studies showing that vitamin E prevented heart disease did not take into account lifestyle factors, such as exercise, said the antiantioxidant crowd. And so on with each study. Still, all these studies might be right on the money, all pointing to the heart of the matter: that we don't understand the intricate relationship between certain types of antioxidants and certain types of free radicals at different moments over the course of one's lifetime. You cannot talk about antioxidants en masse. They all have different potentials. Scientists have been trying to map out these potentials for quite some time now— a vain search, according to some, for a translation of the Babel of tongues that antioxidants and free radicals use in their intricate art of communication. Sir Hans Adolf Krebs won the 1953 Nobel Prize for identifying the free radical–dependent citric-acid cycle, or Krebs cycle, the primary way the body generates energy. The idea that free radicals cause disease—and that antioxidants could prevent this—was first proposed in 1956 in a journal article by Denham Harman, then at the University of California, Berkeley.

Harman is now emeritus professor at the University of Nebraska and still goes into the office every day to keep up on the new antioxidant research. Well into his eighties, he takes his antioxidant supplements every day. He says that free-radical research has come a long way. As Harman relates it, people either ignored or ridiculed his work for the first ten years. It seems that as the "beat-generation" radicals were making their mark in music and literature, "chemical" free radicals remained in the shadows of serious research. By the late 1960s, though, Harmon said he had enough data to show that the average life span of laboratory animals could be increased by decreasing free-radical reactions with antioxidant supplements or diet modulation. By 1972, Harmon said he had evidence that maximum life span was determined by the rate of free-radical damage to the mitochondria.

More scientists became interested in free-radical theory through the 1970s, and they experimented with a variety of antioxidants, each with a different chemical potential for mopping up free radicals. One of the most effective antioxidants, chemically, in a test tube, is phenylbutylnitrone, or PBN. In one famous study, old gerbils given PBN were suddenly able to run through mazes as well as young gerbils. The young gerbils also got the PBN, but the antioxidant didn't affect their maze performance. When the old gerbils stopped taking PBN, they became feeble again and got lost in the mazes. No one to this day knows why. The outcome, unfortunately, has never been replicated. This is the status of the field after forty-some years: Interesting results have been recorded; but scientists have had a tough time replicating them or even explaining the positive results they have found.

Barry Halliwell of the National University in Singapore wrote a short article in *The Lancet* in 2000 entitled "The Antioxidant Paradox." Halliwell lamented the fact that although diets rich in antioxidants seem to have a positive effect on health, popping antioxidant supplements can go either way, and the results are not at all predictable.

If the chemistry is all the same (that is, converting free radicals to neutral molecules), why would a given antioxidant have different effects at different times on different regions of the body? Several entirely different mechanisms might be taking place: Extra amounts of antioxidants might be turning into pro-oxidants, fueling free-radical production and its damage; supplements might do nothing at all because they can't get to where they are most needed; or antioxidants might not be the magic beneficial chemical in the food we eat after all.

Can antioxidants turn on you? Several studies have shown that people who did not get the daily recommended allowance of vitamin C had an increase in free-radical damage to their DNA. But, paradoxically, people who took megadoses of C also had an increase in DNA damage. The second scenario might happen, Halliwell says, because vitamin C can worsen cell damage once it has already started.

Within a cell, certain metal compounds are released as a result of free-radical damage. These metals themselves can act as a catalyst for further free-radical damage when they are in a "reduced" state, with a missing oxygen or an extra hydrogen atom. Antioxidants put metals in this reduced state. Thus, in this environment, antioxidants become pro-oxidants. This was demonstrated in laboratory animals exposed to the pesticide, paraquat, a known carcinogen. Animals that received vitamin C before exposure were more or less protected from cancer. Animals that received vitamin C after exposure—as a sort of medicine—didn't fare as well. The antioxidant aggravated the damage caused by the herbicide and led to more cancers. The American Cancer Society warns cancer patients not to prescribe themselves antioxidants because of this effect.

Compounding the problem is the fact that free radicals can kill cancer cells; that's how cancer treatment works. So taking antioxidants at the wrong time essentially arms the bad guy with the weapons to stay alive and multiply. The same antioxidant that helps a normal cell can help a cancer cell even more, animal studies have shown. No one knows how to get antioxidants to travel where and when they are needed. Most free-radical damage occurs in the mitochondria. The process of making energy in the mitochondria, called the respiratory chain, depends on the availability of free radicals. And extra free radicals are made in the process.

The mitochondrion houses a tiny, circular strain of DNA that contains thirty genes. This is separate from the double-helix DNA in the cell's nucleus. The mitochondrial DNA, called mtDNA, is often the free-radical's first point of attack. When mtDNA is damaged, it cannot do its job of creating the proteins (molecular messengers) needed for daily bodily maintenance. This is where a magic bullet could come in handy: something that could penetrate the mitochondria and mop up a rogue free radical set on doing damage, while steering clear of the intricate respiratory chain. Yet the mitochondrion is a veritable fortress, with a tough outer wall, a motel-like inner barrier, and a meandering inner wall protecting its precious contents. Proteins get out, but antioxidants have a tough time getting in. No one is sure whether brute force, a megadose of antioxidants, is the way into the mitochondria. Perhaps the body has more subtle ways of entering.

Centenarians, those folks who live to age 100, might inherit a gene that provides some internal mechanism for warding off free-radical damage within the mitochondria, and this may help them stave off disease and age more slowly. Some researchers are actually looking for this gene in the mtDNA of volunteer centenarians. Two centenarian studies demonstrated a possible genetic foundation for minimizing free-radical damage and living longer. One study was among Japanese centenarians; the other was of French Caucasians. Researchers found a certain gene structure in the mtDNA that produced a particular protein within the mitochondrial respiratory chain. Centenarians were significantly more likely than noncentenarians to have this unique gene signature in the mtDNA.

It was not clear, however, whether this protein decreased the rate of free-radical production. That's par for the course in this field. The lack of quantitative measurements of radical species has plagued this field for a long time, and no study shows that those who live very long had less oxidation than those who die early.

Thomas Perls of Harvard Medical School, who heads the Centenarian Sibling Pair Study, is also searching for the role of genes in longevity, and he believes that minimizing free-radical damage is a key factor in living to 100 (see chapter 14). Perls says that most of us have the genetic potential to live 85 years and maybe longer. Centenarians may have a gene that slows the aging process, nicknamed the Methuselah gene after the biblical character who lived over 900 years. Fruit flies with this Methuselah gene live 35 percent longer than ones without the gene. Perhaps more interesting, fruit flies with the Methuselah gene also live longer than normal flies when exposed to paraquat, further evidence that free-radical neutralization slows aging. Then again, a scientist can coax a fruit fly to live a little longer through many methods, including chilling them. No Methuselah gene has been found yet in humans.

Think of antioxidants as drugs. Would you take a drug that hasn't been proven to be safe or effective at certain doses? Many doctors see no harm in taking supplements, as long as the dose is not too high. Nearly all doctors will agree, however, that exercise and diet constitute the master therapy. A varied diet seems to be more healthy than simple supplement taking because the isolated antioxidant in the capsule might not be the superhero. Fruits and vegetables are rich in antioxidants, but these plants contain hundreds of other chemicals. Any single chemical or combination of chemicals might pack the therapeutic punch.

Nutrients from food enable the body to make its own antioxidants. A chemical produced by the body called glutathione is ultimately responsible for neutralizing free radicals; and the glutathione concentration in cells dwarfs that of the free-radical scavengers such as vitamins C and E. Diet and energy demands determine the amount of free-radical generation and removal, with supplements playing a minuscule role, if any. The production of free radicals, absent genetic defects, results from normal metabolic processes; and the destruction of free radicals in a nonharmful manner is also the result of normal metabolic processes.

We saw how beta-carotene supplements were deadly for smokers. SOD, or superoxide dismutase, an enzyme billed as the most powerful antioxidant known to humankind, is another useless supplement. In a pill form, SOD breaks apart when digested. SOD is an important enzyme, but only when the body produces it on its own. Any health-food store clerk who tells you otherwise is lying or ignorant.

Vitamin E is a funny thing. A few doctors are still excited about vitamin E, which is found naturally in vegetable oils (particularly wheat-germ oil), sweet potatoes, avocados, nuts, sunflower seeds, and soybeans. But alas, support is waning. One theory is that the oxidation of low-density lipoprotein (LDL, the bad cholesterol) is the first step in plaque formation in arteries. Vitamin E might inhibit this oxidation, thus reducing the risk of atherosclerosis and heart attacks. The problem has been finding a study to support this theory; all of the studies—and some have been huge—have struck out. The perhaps aptly named CHAOS (Cambridge Heart Antioxidant Study) found that high doses of vitamin E lowered the risk of a second heart attack but raised the risk of dying from that second heart attack if it came. The big Italian GISSI-Prevenzione study and the American HOPE study saw no effect from vitamin E either way in preventing heart disease.

Vitamin E may also cause bleeding problems, particularly in people taking anticlotting medications. By late 2001, studies were showing that antioxidants—and quite likely vitamin E—were hindering the benefits of cholesterol-lowering drugs called statins. Whatever your position on America's dependency on pharmaceuticals, statins have clearly saved millions of lives. Antioxidants have not been shown to save any lives. And now they get in the way of statins.

The American Heart Association doesn't recommend antioxidants. The American Cancer Society doesn't recommend antioxidants. The National Institutes of Health don't recommend antioxidants. To quote Richard Veech, Chief of the Laboratory of Membrane Biochemistry at the National Institute on Alcohol Abuse and Alcoholism, who has reported on the interplay of free radicals and antioxidants for over thirty years: "People don't want to exercise. They don't want to eat healthy food. They don't want to stop drinking; they don't want to stop smoking; they don't want to stop having dangerous sex. They want to take a pill. Well, good luck."

22

𐕣 𐕣 𐕣 𐕣 𐕣 𐕣

The Unbearable Heaviness of Being: Fat People and Food

People come in all types of shapes and sizes. There's no myth here. In the United States, however, people seem to be gravitating toward one shape: round. Now is not the time to vilify the overweight—nor the skinny, for that matter. Now is the time for us to admit that Americans are heavier now than at any time in history, that this is unhealthy, and that we need to lose weight.

The National Institutes of Health estimate that over 60 percent of America is too heavy, and that number could soon reach over 90 percent, because the rise in obesity is particularly acute among children. There's no conspiracy to mandate a Hollywood ideal of thinness. There's no denying that many people are voluptuous or stocky by design. The problem is that beautiful voluptuousness is turning into unhealthy fat because of changes in our lifestyle and diet. Health experts are merely suggesting we maintain the kind of weight that our ancestors—until about fifty years ago—maintained. This is an issue of objective health, not subjective beauty.

Yes, lots of skinny people never put on much weight no matter what they eat. Others, albeit very few, pack on the pounds at an unnatural rate. The vast majority of us are in the middle. If we consume more calories than we burn off through exercise and daily metabolism, we gain weight. This is exactly what is happening. It is only in recent history that humans (and their house pets) have joined the ranks of livestock as being the only animals that consume more calories than they burn. We aren't any different biologically from our ancestors. We aren't any lazier. The problem is that we are working hard in a different, less physical way. And we are eating foods made solely of the unholy trinity: fat, salt, and sugar. We eat high-calorie foods that are harder to burn, and we physically burn fewer calories. Thus, we are overweight. This is natural.

As more people enter into the realm of chubbiness, more tricksters try to sell their "bad medicine" diets—such as the all-protein diet, quite possibly the most ridiculous and irresponsible of them all. A heavy nation is not a healthy nation. Being obese and overweight are major risk factors for circulatory disease, diabetes, and cancer, our main killers. This is why the NIH uses the term "epidemic" when talking about the national heft.

Obesity is defined as 20 percent or more above ideal weight; being overweight is a matter of being a few pounds over the thirty-pound weight range based on sex and height. The standards defining overweight are admittedly flawed. There are folks who will naturally carry a few more pounds over the standard and remain healthy. Such folks, however, usually have maintained this "extra" weight throughout most of their lives. A beer gut earned in your thirties doesn't count as natural stockiness. Obesity should never be confused with body shape. Barring a thyroid or metabolism disorder, which are rare, no one is obese by design.

Never before have so many been so well fed. Yankee Stadium, built in the 1920s, has had to remove 9,000 seats and increase seat sizes from 15 to 19 inches to accommodate the modern American rump. Thank our fantastic food supply and fantastic conveniences. The one-two punch: we mass-produce and readily consume the most fattening of foods, such as dairy products, meats, fast foods, and prepared foods, and we have created a society in which we hardly need to expend calories—with cars instead of sidewalks; escalators and elevators instead of steps; video games instead of stickball; power tools instead of manual ones; garage-door openers, rarely seen twenty years ago; houses and neighborhoods geared toward minimizing all physical activity. Every opportunity we have to physically move our bodies from one place to another is being replaced by technology. Even pencil sharpeners are electric. "Oh, my arm. Is there no relief from this drudgery of sharpening a pencil mechanically?"

Weight gain is a natural response to this American lifestyle. It's just too darn easy to put on pounds. We aren't bad people. It's just that we've set up a system of wonderful conveniences whereby we have to go out of our way to exercise. It takes great will and a bit of opportunity, given the abundance of our blessings, to ride a bike five or ten miles to work or to take time out of our days to devote to exercise—something our ancestors never had to do. You cannot expect half of Americans *not* to be overweight. Thus, one should not be ashamed of being fat. For a good 90 percent of us, we have to work extra hard—break the American lifestyle—in order *not* to gain weight. As for the obese, they are not necessarily gluttonous pigs stuffing their faces with cupcakes and refusing to exercise. The heaviest among us have most likely arrived at that weight through a combination of little exercise and a series of bad diets that have ravaged their metabolism rate and left many of them eating a minimal amount of calories for survival yet still gaining weight.

Weight gain is complicated. The first myth of weight gain is the idea that you were destined to be fat because you inherited the "fat gene," which leaves you no option other than to be rotund. Thus, the search for the fat gene will ultimately lead to a world of skinny people. No. Very few people—less then a hundredth of a percent— are obese because of a malfunctioning thyroid or hypothalamus or a genetic disorder. Rarely can you truthfully say to yourself: "I'm obese, and it is natural for my body to be this way." Go to Africa for a few years, walk twelve miles a day for water, and live on millet and locusts. You'll lose weight. Likewise, no nation or ethnic group is genetically excused from obesity. The Asian diet—very little meat and ample amounts of vegetables—is what keeps Asians thin. Asians in America grow chubby with the rest of America. Asians in Asia, in fact, are growing ever more corpulent as they switch to the American diet and lifestyle. The "fat gene" argument only goes as far as suggesting that certain people need fewer calories than others do and, as a result, may have a normal weight range of ten to twenty pounds more than others—not a hundred to two hundred pounds more.

The second myth of weight gain is that diets work. No diet works. The National Institutes of Health estimate that 95 to 98 percent of diets fail to keep off weight for more than three years, and over 90 percent of diets *lead to further weight gain.* Even the most straightforward of diets—simply cutting back calories—won't lead to weight loss. The only way to lose weight and to keep it off is through a change in lifestyle.

WHY YOU WEREN'T MEANT TO BE FAT

Let's examine the first myth, that we were meant to be fat. If that were true, then the same percentage of the population that is overweight now would have been overweight way back when, say 100, 500 or 1,000 years ago. Go to the footage. Take a look at a crowd watching a baseball game in the 1920s. Take a look at the average weight. You will find a fat person, to be sure; but you won't find that every other person is fat. This is the way people really were. The grainy film doesn't lie. Reports through the centuries support this anecdotal observation. Africans, Asians, and Aztecs all saw obesity as a rare event, a result of supernatural forces, not overeating. These societies often raised the occasional corpulent individual to the level of seer. These obese most likely suffered from abnormal metabolism. In medieval Europe, as in ancient Rome, most obesity was a result of overconsumption and inactivity among the wealthy. Obesity, corpulence, and other terms for degrees of heft have never been well defined until recent times. Nonetheless, statistics from the nineteenth century show that fewer than 5 percent of Americans were obese by today's standards. The heaviest Americans then were the wealthiest, the so-called fat cats. Since the 1960s, U.S. obesity rates have surged from 5 to 10 percent to 12 to 50 percent, depending on the population. Today, the rich are often thinner than the poor and the middle classes.

You can't go by Hollywood standards, however, to see how thin most people used to be. Years ago, chubby girls were featured in movies and snapshots because they were considered to be attractive; they were the rarity—a symbol of American prosperity. Today skinny girls are in (and they are maligned as unhealthy). Can you imagine people 100 years from now assuming that all of

America was skinny, judging from the Hollywood movies of today? You can't judge by old painted portraits, either. Artists, upon request of their rich customers, actually added a paunch or cushy fat around the limbs to show that the subject was living a life of leisure, free from the daily toil that kept others thin. Their clientele paid to look fat, even if they weren't. The eighteenth-century painter John Singleton Copley was particularly adept at portraying lean patrons with flattering adipose.

The vast majority of today's overweight Americans would have had a healthy, lower weight five hundred years ago. Life was filled with cutting, lifting, hauling, washing, walking, and constant, constant effort that burned off calories. Their lives of toil weren't necessarily a good thing, either; they often led to exhaustion and an early death. Nonetheless, people were thin. Rest assured, you would have been thin, too—same genes, same person; different era, different weight. Life was rigorous and food was lean.

"Oh, to be fat!" is what most people would have said a few hundred years ago. There was little concept of dieting, for the food that the common folk ate was scarce and unfattening. Most people ate vegetable soup and gruel, a sort of mushy mix of grains in water or watery milk. Famine was a constant threat. Rarely was there meat to eat, and even more rarely was meat fatty. An Italian painting from the 1500s depicting Utopia, a heaven on earth, shows roasted chickens raining down from the sky. That's how rare meat was. Talk to some of the old folks in America's Chinatowns, and they'll tell you how they only had meat once or twice a year in China, during a festival. Their grandkids in America have meat every day. And the kids are stout. (The grandparents are often happy about it because fat kids are a sign of health and wealth in the Chinese culture.) The twentieth century brought an influx of fatty foods to the wealthier nations, and our bodies weren't ready for it. They still aren't; fat foods make fat people.

WHY DIETS DON'T WORK

Now we are getting to the heart of the reason that diets don't work. The body doesn't like losing weight. With all this talk about lean times throughout history, you can imagine that the body tries to cling to as much fat as possible. Fat is long-term fuel, the substance that kept the caveman alive during days or weeks without food. Our modern-day body craves fatty food, thinking that a famine is just around the corner. After all, we are only a few thousand years out of prehistory. In evolutionary terms, that's nothing. Our bodies are essentially the same as those of the early humans.

So, you put on a few pounds: you indulge your natural craving for fatty foods (they taste so good) and you neglect to haul ten pounds of water for eighteen miles to burn it off. The next step is a diet, to restrict your calories and lose weight. The body reacts to such a diet as if it were starving and reduces the calories it burns. Your body also doesn't want to give up that fat too quickly because it doesn't know when the famine will end—or when the next one will begin. If the body takes in any more fat, it will hold on tightly to this precious commodity.

Your body enters "calorie conservation" mode. To lose weight, you will have to cut your caloric intake even more and stay at that lower level. That is, now you must eat a *lot* less if you want to lose pounds and avoid hanging out at your current weight, because your body has reset itself at a new rate of metabolism—a slower rate of turning food into energy, which requires fewer calories. Consider an example of two women, each weighing 130 pounds. One woman once weighed 145 pounds but lost 15. The other has always been 130 pounds. The once-heavier woman's metabolism has reset itself to burn calories more slowly as a reaction to losing 15 pounds. This woman must now eat 250 fewer calories each day than the woman who has always weighed 130 pounds just to stay at that weight. Doesn't seem fair, does it?

Diets that restrict calories can technically work, but it's hard to eat a lot less to maintain a given weight. If you goof and eat a "normal" amount of food, you will gain weight. You enter a period of eating no extra food and yet you gain weight. The pounds pile on. To maintain this heavier weight, you have to eat even less than the last "less." Now you are at the point of eating much less and still not losing weight, just staying even. Goof up enough times and you will reach the point where you are dieting— eating less—and still gaining weight, all because your body, fearing starvation, has slowed down your metabolism. Soon you will have

to eat even less than the minimal amount of food needed to nour-
ish yourself (around 900 calories a day) or continue gaining
weight. Many obese dieters are at this stage. Dieting truly is a losing battle. There is little room for error. You must be incredibly
disciplined.

Exercise helps because you can burn off 250 calories instead of
"not eating" 250 calories. In that first scenario, the woman who
lost 15 pounds to get to 130 pounds will not enter into "starvation
mode" if she exercises. By exercising, she doesn't forfeit food and
her body assumes that all is normal. All is very normal, actually,
because working up a sweat and burning calories is natural, as far
as the body is concerned. Her metabolism stays high. Thus, the
best advice is to live like a caveman and burn off as many calories
as possible each day through physical activity.

The other trick to the weight loss game is to never gain weight.
This is maintained not only through diet but also through lifestyle.
Chinese peasantry is a lifestyle, albeit a miserable one, that will
keep you thin. Franciscan monks also tend to be slim from all that
gardening and vegetable eating. Do you need to go to such ex-
tremes? Probably not. Lifestyle means diet combined with exercise
incorporated in such a way that it is not a strain but rather a natural way of life. The Pritikin diet, for example, stresses a lifestyle
of nearly no fat and plenty of casual exercise, like walking. Meat
is allowed, but only in very small, nonfat quantities. The Pritikin
diet has proven to be rather successful not only in keeping people
slim but also in being "gentle" enough that most people can follow
the lifestyle without feeling like fasting monks. The Japanese lifestyle of little meat, some fish, and plenty of rice and vegetables, including sea vegetables, incorporated with biking and walking is
another lifestyle that many Americans can adopt easily. (Sadly, Japan
is adopting an American lifestyle of pork and beef and few vegetables, and the population is slowing growing heavier as a result.)

These lean lifestyles work best for adults already at an ideal
weight or slightly above it. People who are very heavy usually put
on the last of those pounds through dieting. Losing hundreds of
pounds is not impossible, but it is pert' near impossible. Many
argue that weight fluctuation (up and down, up and down and up)

is more unhealthy than maintaining a high weight, say 200 pounds. There is a grain of truth to that, as we shall see below.

EAT THE FOODS YOU LOVE
AND STILL LOSE WEIGHT

You have probably come across lots of crazy diets. Whenever you hear "Eat the foods you like and still lose weight," just run. Running, in fact, will help you lose weight better than whatever such a diet recommends.

Many diet programs, such as Weight Watchers, are concerned with calorie counting, which serves as a daily reminder, on paper, of your sacrifice and how little progress you will make by merely restricting calories. The calorie-counting culture has produced marketing, such as the Tic-Tacs campaign, which bills its product as the one-and-a-half-calorie mint. Other mints have, heaven forbid, four or five calories. Of course, they are four or five times bigger than the tiny Tic-Tac, but that logic seems to be lost. You can make a one-and-half-calorie cake, after all, if you make it the size of a crumb. Regardless, the difference between one and four calories cannot really be measured. And the mere act of taking the wrapper off the mint and lifting it into your mouth is probably burning off those calories. I wonder which mints are better to pop after eating that Big Mac, large fries, and 128-ounce soda?

My favorite silly diet is the Atkins diet. This is the one that says you can eat all the bacon, fatty pork chops, and cheeseburgers you want and still lose weight. The Atkins "all protein" diet is a unique combination of the irresponsible, the illogical, the incorrect, and the harmful. Other diets are usually only one or two of these. The Atkins diet is very popular in America because it plays into the American philosophy that you can reach your goal (physical, financial, whatever) without doing any work.

Robert Atkins's premise is that carbohydrates, not fat, make you fat. He states this clearly in the beginning of his top-selling book, *Dr. Atkins' Diet Revolution*. Here's the gist: obesity is a result of whacked metabolism, and obese people gain weight on fewer calories than thin people. (This much is true.) Carbohydrates raise the amount of glucose in the blood, called blood sugar, and trigger the pancreas to secrete insulin. (He's still on solid ground.) A diet high in carbohydrates leads to hyperinsulinism, or too much insulin, which ultimately compromises the body's ability to use insulin to metabolize glucose and regulate energy consumption, elimination, and weight. (The theory starts to flounder. Carbohydrates per se aren't the culprit; overeating is.) A diet with hardly any carbohydrates and more protein would set the insulin-metabolism process in order. (Not quite.) A diet with carbohydrates as the staple is unhealthy; humans, from the beginning, ate mostly meat and remained robust. (Say what?)

By carbohydrates we mean grains, such as rice and wheat, and most vegetables. Are carbohydrates unhealthy? Not at all. That's a preposterous idea. The entire world outside America eats carbohydrate-heavy meals, not meat-based meals, and the entire world outside of America is largely slim—or at least had been until McDonald's and other Americanisms moved in. Rice is the staple food for billions of people. The healthiest meals are made predominantly of vegetables and carbohydrates (rice, couscous, tortilla) and very little, if any, protein from meat.

To suggest that early humans not only survived but thrived on meat, as Atkins does, is equally preposterous. Securing meat thousands of years ago was more difficult than driving to the supermarket to purchase prekilled, cut, boned, packaged, refrigerated strips of unnaturally fattened chicken, beef, and pork. Think about it: you're naked and in the woods. Now get something to eat. Early humans scavenged for whatever they could find. The human body, in fact, has been incredibly resourceful in its ability to survive on a variety of foods—mostly roots, seeds, and leafy green vegetables. Hunting the mammoth may seem glamorous, but it was very hard. Native Americans in the Great Plains area of the United States didn't dine on bison every night. The bison hunt came only a few times a year. Yes, they loaded up on this meat, but then they went back to good ol' corn, beans, and squash during the rest of the year. The natives of northern Canada, commonly called Eskimos, were one of the few meat-eating societies, chiefly because they couldn't grow vegetables. When they couldn't catch food, they went hungry. These people died primarily in two ways: they were eaten by polar bears or they starved. Not a fun life. I challenge Robert Atkins to survive by catching his own meat before boasting how natural meat eating is.

The advent of grain cultivation was a milestone in history. Humans, for the first time, could store grain (those nasty carbohydrates) for years to use during times of famine. Fewer people starved. More people lived longer, healthier lives—on carbohydrates, not meat. In fact, all of civilization is based on grain. Grain became a commodity, an element of wealth. Cities developed by virtue of grain harvests.

In addition, eating predominantly carbohydrates does not lead to insulin problems. Overeating does. Atkins suggests that the high rate of type II diabetes in America is caused by eating carbohydrates; once people have diabetes they gain weight because insulin can no longer regulate metabolism properly. Actually, the reverse is true. People who are overweight—from a sedentary life and a diet filled with fattening foods, including carbohydrates but also fried pork chops—develop type II diabetes. This is the hyperinsulinism Atkins refers to in his book, an excess of insulin secreted by the pancreas from all the food entering the digestive system. It is the process of gaining weight that puts one's metabolism out of whack, not eating carbohydrates.

A civilization based on supplying meat to the legions of Atkins dieters is frightening indeed. Vast areas of land must be cleared to grow the grain that feeds the livestock. A farmer can produce twenty times more protein per acre by planting soybeans than by planting cow food to fatten a cow for protein. At this very moment, forestland is being destroyed in Brazil to create barren grazing fields for the sole purpose of growing fast-food burgers. Billions of tons of manure produced annually from American cattle release methane gas into the atmosphere, adding to greenhouse warming. The mass production of meat is irresponsible to the environment. The planet could never sustain a world of Atkins dieters.

Mind you, there's nothing wrong with protein, aside from its being overrated. (Few people die from protein deficiency.) It wouldn't be so bad if Atkins were pushing protein from beans.

Instead, he is pushing beef and pork, which are riddled with fat. The cavemen—those inveterate meat eaters, according to Atkins— never encountered anything so unnatural as beef and pork. The game they hunted were naturally lean: antelope, wild fowl, and insects, to name a few. Hogs and cattle are a modern-day invention. The consumption of such fattened food in high amounts is intimately linked to stroke, heart attacks, and many cancers.

The eerie part about the Atkins diet is that it does help shed pounds quickly in the short term. The effect is the same as starvation: the body starts burning fat because there are no carbohydrates to use for fuel. Yet after a couple of weeks, a process called ketosis kicks in, which is the accumulation of ketones, acids such as acetone created as byproducts of burning fat. Ketosis can be dangerous; runaway ketosis leads to brain dysfunction and coma. After a couple of weeks, Atkins recommends eating more than just meat by taking vitamin supplements (so much for nature's diet), and adding a few vegetables to your meals. How much ketosis is too much ketosis? You'll have to read the Atkins book. You can test ketone levels in your urine. Be your own doctor. You have only brain dysfunction and coma to fear.

The long-term effects of such a meat-heavy diet are not good. The excess fat from the meat ultimately raises cholesterol levels and deposits fatty gumdrops in your arteries' walls, which constrict the flow of blood and lead to stroke and heart attacks. The lack of nutrients from the dearth of vegetables in the Atkins diet can lead to all sorts of problems, from poor skin and hair loss to chronic infections. (Atkins recommends a "Dieter's Formula" of thirty-one vitamins and minerals in pill form; if only the cavemen had a CVS.) And watch out for the constipation, fluid retention (take asparagus tablets, but God forbid, don't eat asparagus), fatigue, insomnia, and other potential nuisances listed at the end of Atkins's book. The bottom line is this: Atkins has been at this game for over thirty years. In that time, he has yet to publish a peer-reviewed article showing the benefits of his diet compared to other diets in a clinical setting. Purveyors of bad medicine publish anecdotal evidence of success in their own books. Purveyors of good medicine publish the results of real studies in *The New England Journal of Medicine, The Lancet,* and the like.

OBESITY RIGHTS

Obese individuals endure broad discrimination—from dirty looks and assumptions of slobbery to lost jobs and difficulty adopting children. The National Association to Advance Fat Acceptance (NAAFA) does a wonderful job in advocating obesity rights. The group argues that fat people can be fit, which is true, particularly for large athletes, and that being heavy and maintaining that weight (instead of dieting and fluctuating) is as healthy a lifestyle as being slim. This much is also true, provided that the excess weight level isn't too high. NAAFA supports this idea, however, with studies that show that obese people in cultures that do not discriminate against fat people are healthier than obese people in America. The theory is that obese people in fat-friendly cultures do not encounter the stress, guilt, discrimination, and diet-induced weight fluctuation that contribute to poor health.

This argument is flawed. Polynesians, often considered naturally fat, were actually stocky and muscular before encountering Europeans. These cultures are accepting of fat, and many folks living on South Pacific islands are indeed fat. Yet, island nations such as Tonga and the American state of Hawaii consider obesity their prime health problem. Diabetes, once nonexistent, has consumed Polynesian populations at an alarming rate. Even children are developing type II diabetes, which leads to circulatory problems, poor vision, and, quite often, early death. Likewise, the once-stocky Eskimos—the Inuit tribes of Canada and Greenland—have grown soft and unhealthy from the influx of soda pop, prepared foods, and a lack of exercise. They pay the price (as accepting as they may be of their fat neighbors) with poor health and suicide. Obesity, diabetes, and depression are rampant among Native American cultures. These cultures across the United States and Canada are relatively insulated from the Hollywood ideal of beauty. Still, they are collectively unhappy about their obesity. The

same goes for the aborigines of Australia, where the adult incidence of obesity in the Torres Strait islands at the northern tip of
Australia is approaching 50 percent. It would be a tragedy for
Native Americans and Australians—and an insult to their cultures—to accept the fact that they are fat, for fat represents the
cultural oppression they have endured over the last two hundred
years. Fat, for the vast majority of people, is simply unhealthy.

CARE FOR SOME CRICKETS?

The word *diet* comes from the Greek *diaita,* meaning a prescribed
way of living. These days, unfortunately, *diet* refers to a gimmick
to lose weight quickly. Perhaps we should get back to that original
meaning and change our way of living. Clearly something is wrong
with America, because skinny people come here and get fat. It's
not the water. It's not the carbohydrates. It's not genetics. It's not
even that Americans are lazy, for we are sleeping less, working
harder, and juggling more tasks than at any time in history. The
problem is the fatty, prepared foods and the lack of physical movement. The search for the obesity gene or diet drugs is simply the
search for a way to maintain ideal weight while continuing to be
inactive (toss that remote control over here, honey) and eating
fatty processed foods.

I'm not suggesting a bag of dried locust or crickets, a popular
snack in southeast Asia. We simply have to evaluate the fact that
the two most popular vegetables in America are potato chips and
french fries. Clearly, some of us put on the pounds more easily
than others. We cannot be fatalistic in our thinking, however, that
weight gain is inevitable. In another time and another place, we
would have been slim.

23

Not Milk? Milk and Your Health

Milk. Most cultures don't drink it, and most people can't digest it. Milk's claim to fame is calcium, the mineral known for making bones strong. Milk is loaded with calcium, and calcium is rather important. Yet milk is also loaded with fat, animal protein, and artificial hormones injected into cows to enable the poor creatures to produce twice the amount of milk that their bovine ancestors did just a century ago. No one is sure how healthy this other stuff is or how it affects the level of calcium that is actually absorbed by the body. It is clear, however, that calcium-rich vegetables are much better than milk for the body.

What to do, what to do. Calcium is so important, but milk comes with such baggage. The obvious solution would be to get calcium from other sources; that's what most of the world does. There's collard greens, for example. Calorie for calorie, they have far more calcium than milk. But alas, who eats collard greens or any other calcium-rich leafy green vegetables? Then there are sardines and anchovies, tofu and broccoli, chicken cartilage and beans. . . . Not to your taste? I'll stop here.

The major health organizations in America, both governmental and private, advocate drinking milk for its calcium content, and dairy products have retained a cherished block of real estate in the food pyramid. Some of the more paranoid among us accuse the multibillion-dollar dairy industry of having its way with these health organizations. This may or may not be true, but health experts in the United States have little else to advocate for calcium other than milk. No one will pose with an anchovy mustache. The last thing the health pros can do is tell people—especially kids—not to drink milk when the alternatives are soda and sugary drinks.

What's good about calcium? Yes, strong bones. Yet this vital mineral is not just locked up in bones for life's long haul. Calcium flows through the bloodstream and is necessary for muscle contractions, steady heartbeats, and transmitting nerve impulses. Calcium is also key in energy metabolism and waste elimination. Bones continuously release calcium into the bloodstream for these functions and soak up new calcium from foods such as anchovies. Adolescents need the most calcium, for their bones, muscles, and nerves are growing the fastest. Up until about age thirty, we can store much of the calcium we consume in our bones. The calcium deposit is important because, later on in life, bones cannot absorb as much calcium as they lose. Calcium reserves from our youth, like a pension plan, support much of the daily calcium needs of old age. Without a solid reserve—or without a constant influx of new calcium—bones can become weak and easily break. This is why old folks need calcium just about as much as teenagers do. Osteoporosis is a disease in which the bones, for reasons unknown, release *far* more calcium than they absorb, an out-of-control calcium leeching. Many postmenopausal women in America suffer from some degree of osteoporosis. Some men get it, too.

What's bad about milk? First, the fat. When the National Institutes of Health or the National Osteoporosis Foundation tell you to drink milk, they mean the nonfat kind, which isn't so popular. Whole milk is 4 percent fat, and milk fat is very good at making you fat. That's the purpose of mother's milk, to make a baby fat. U.S. prisoners of war in Germany and Japan were given ice cream and milk fat after being liberated to fatten them up before going home. That's fine for a couple of weeks, but with a lifetime consumption of milk fat comes a lifetime accumulation of body fat. Fat raises cholesterol levels in the blood and leads to clogged arteries, strokes, and heart attacks. If you want your milk to be healthy, you'll at least have to drink the nonfat variety. True, kids drank milk in earnest in the 1950s with no apparent ill effects (aside from the argument that middle-age America today has high cholesterol levels). Sadly, kids do not play now the way kids played then to burn off calories, and the fat starts accumulating early these days.

Milk also has animal protein, which isn't bad per se. The funny thing about animal protein, though, is that it triggers the release of calcium from the bones through the body and down the toilet. So, the more milk you drink, the more calcium you lose. Scientists argue over the exact ratio, with most believing you get a net benefit of calcium from drinking milk. Some, however, point to the fact that countries with the highest dairy consumption—Scandinavia and the United States—have the highest rates of hip and other bone fractures, a common measure of osteoporosis. Researchers at Yale University, in fact, have identified a worldwide correlation between animal protein consumption in general and osteoporosis—even to the extent that osteoporosis is rare in South Africa yet common among milk-drinking, meat-eating African Americans. Conversely, the ongoing, multiyear, Harvard-based Nurse's Health Study has found no evidence that milk prevents hip fractures in older women, in contrast to those "Got Milk?" commercials.

Excess calcium seems to eliminate a certain form of cancer-fighting vitamin D in the bloodstream as well. Ironically, vitamin D is needed to fuse calcium into the bones. Normally our bodies generate plenty of vitamin D, indirectly from exposure to sunshine. Sun-starved Scandinavians can be deficient in vitamin D, particularly in the wintertime, so they are faced with a triple-whammy of leached calcium entering the bloodstream, low levels of vitamin D, and even lower levels of the cancer-fighting type of vitamin D.

Milk is essentially liquid meat, and because of the leaching issues, America's high-protein diet from meat and dairy pumps up the daily calcium requirement. Americans need to get 1,000 to 1,300 milligrams of calcium a day. The protein in/calcium out connection is well known. Asians, in stark contrast, can get by on 500 milligrams or less a day and have stronger bones to show for it. They eat less meat and get calcium from leafy green vegetables, tofu, and small fish with edible bones. Asian cultures have much lower rates of osteoporosis compared to the rest of the world and rarely eat dairy products, at least traditionally. Japan has the highest rate of dairy consumption in Asia and the highest rate of osteoporosis. Calcium leaching aside, milk might not deliver all the calcium it promises, despite its high-calcium content, for it is all a matter of absorption. The body absorbs only 32 percent of milk's calcium, compared to over 50 percent of the calcium from kale, broccoli, mustard greens, turnip greens, and brussels sprouts. So, calorie for calorie, whole milk is one of the worst sources of calcium. If you're going to drink milk, you would best benefit from skim milk.

Hormones in milk is another sticky issue. Mother's milk, by design, delivers the nutrients and antibodies a baby needs to build its immune system. But the poisons of the mother are also passed on to the infant. Toxins from cigarette smoke easily make their way into mother's milk. So, too, does alcohol; you can get a baby drunk with whiskey-tainted mother's milk. Cows and humans aren't that different when it comes to milk production. Animal antibiotics and human-manufactured growth hormones that are injected into the cow are concentrated in milk, and we drink it. This is different from pesticides sprayed on food, which is not a big health issue. Pesticides can be washed off; antibiotics and hormones are concentrated within the food. One hormone made by Monsanto called recombinant bovine growth hormone, or rBGH, may cause cancer; studies are ongoing. The European Union is against rBGH. These countries placed a two-year ban on it in 1994 and have renewed the ban through 2002. True, Europe is against lots of good things and in favor of lots of dumb things, like homeopathy and David Hasselhoff's singing. Still, they are worried enough to spend a few million dollars to investigate the issue.

The rBGH increases milk output. Cows injected with rBGH usually die early or develop udder infections, which leave their milk filled with pus. But this is an animal-rights issue outside the context of human health. The human health consequences of rBGH are unknown. This hormone seems to trigger elevated levels of another bovine hormone called IGF-1, which milk consumers then drink. A Harvard study of 15,000 men published in the journal *Science* in 1998 found that those with elevated levels of IGF-1 in their blood were four times more likely to get prostate cancer. This is the one big study that antimilk and anti-Monsanto people latch on to. Sure, just one study doesn't prove anything. Yet Monsanto is behind rBGH, which makes people nervous because of this company's track record of hiding its activities. Remember Monsanto's "Without chemicals, life itself would be impossible" slogan?

Lots of cows get rBGH, and millions of people drink their milk without knowing this fact. Until 1997, there was a strange law that prohibited dairy farmers who proudly *didn't* use rBGH from bragging about this fact on their products' labels. The makers of Ben & Jerry's ice cream were also restricted from telling their customers they didn't use milk from cows given rBGH. In a classic case of crying over spilled milk, Monsanto's attempt to soften a damning television report about the company backfired. Two news reporters from a Fox Television affiliate in Florida, Jane Akre and Steve Wilson, tried to report about Monsanto and rBGH in 1996. Monsanto pressured the affiliate not to air the story. The station capitulated and eventually fired the reporters. Akre successfully sued Fox for violating Florida's whistle-blower law. And Akre and Wilson won the 2001 international Goldman Environmental Prize for their work, with their faces adorning a full-page ad in the *New York Times* explaining the incident.

But forget all the politics. Let's examine the fact that 75 percent of the world's population is lactose intolerant, meaning they lack the enzyme needed to comfortably digest milk. They can drink it; they won't die. They simply experience stomach cramps, soft stools, and flatulence. It's kind of telling that milk is white, because white people of western European descent are about the only ones who can handle drinking it as adults. All infants can produce the lactose enzyme, but most lose this ability after weaning. According to a 1996 study in the *Journal of the American Dietetic Association,* perhaps as many as 50 percent of Mexicans, 70 percent of African Americans and Native Americans, and up to 90 percent of Asian Americans are lactose intolerant. Got lactose? These folks can handle only a little milk at a time. Europeans, the first to experiment with cow's milk, developed the lactose enzyme over the last ten thousand years or so. Drinking milk is relatively new for humans.

Dr. Benjamin Spock, America's best-known pediatrician, came out against milk after becoming a vegetarian in 1991. He advised that no child older than two years should drink milk. The medical establishment lashed out against Spock, saying that milk contains crucial nutrients for a growing child: calcium, riboflavin, and vitamins A and D. This is true. Soda and other sugar drinks have none of this. But then again, riboflavin and vitamins are added to milk; milk doesn't contain them naturally. We can add vitamins to any drink, really. Orange juice now comes fortified with as much calcium as milk. You can drink that. Fifty years ago, orange juice wasn't so popular. This is why schools mandated the addition of *fortified* milk to school lunches. What else was there to drink?

Expect the milk industry to continue with its slick "Got Milk?" and milk-mustache ad campaigns. And be happy if your kid is drinking milk (if it is nonfat and drug-free) instead of cola. But when you see that old slogan, "Milk does a body good," perhaps think of the more accurate albeit less prosaic battle cry: "Milk might do your body better than some of the other crap you can drink, but there's no scientific evidence to prove this, and your body probably can't digest it."

24

弘 弘 弘 弘 弘 弘 弘

Organic Reasoning:
The Benefits of Organic Food

What does the word "organic" mean to you? For this city boy, organic used to mean from the local farm. I envisioned weathered farmers in their big straw hats driving beat-up tractors across their meager patches of land in the bucolic, rolling hills north of the city. Cows that produced organic milk were happy cows, I thought, dancing in green fields and jumping over the moon, just like the drawing on the bottle. Organic chickens slept under a warm blanket of stars, willingly placing their heads on the tree-stump chopping block when their sweet, natural life drew to a close.

Not so. Organic is big business with mass consolidation. A company named Horizon, for example, controls nearly 70 percent of the organic milk market from its Colorado base. Many of the cows producing organic milk are as cooped up as the cow-robots producing conventional milk, never seeing the light of day, milked three times daily. The only difference is that they are fed a bin of organic feed. The same can be said for organic chickens, even if the label says "free range," for the birds can still be penned in, free to range with thousands of their brethren. (Some farmers remove the chickens' beaks so they don't peck each other to death in such tight quarters.) This isn't always the case, but you cannot rely solely on the organic label for peace of mind. The same holds true for vegetables. In California, five farms grow half of the country's organic food, often right beside conventional crops. Each year, more local organic farmers must sell out to the "organic" corporations, which grow organic and conventional food side by side. Once a farm grows large, consumers have greater difficulty determining whether its food is truly organic—fertilized with "organic" manure from cows that grazed on organic grains, and free from chemical runoff and spray from the conventional farms just next door.

What *does* "organic" mean? It's a term that makes chemists giggle, that's for sure, as all food is organic. Technically, organic refers to something that contains a chain of hydrogen and carbon atoms, called hydrocarbons. All living organisms are organic. So is gasoline, because it comes from decayed organic matter millions of years old. A dry cleaner in my neighborhood brags that he uses only "organic solvents." Clearly he wants to capitalize on the public perception that "organic" equals "safe." Truth be told, every dry-cleaning solvent since the creation of the technique in nineteenth-century France has been organic. Today, the solvent used by over 85 percent of dry cleaners in the United States is perchloroethylene, or perc for short. Nasty stuff. Totally organic. Rocks are inorganic.

The "organic" definition used by the counterculture and back-to-the-land movement in the early 1970s did mean something. This was the dawn of the organic movement. Organic farmers back then used fertilizers of animal and plant origin, commonly called manure and compost, and natural pesticides called ducks and wasps to eat weeds and plant pests. This is the meaning of organic today as referred to on current food labels—not the food itself but the process used to cultivate it. The early organic farmers grew a variety of crops in small quantities to keep the bugs at bay, for too much of one type of crop—say, corn—would attract corn bugs. The movement was certainly legitimate, and the legacy continues today. Many of these renegade organic farmers have survived, albeit with meager profits, by selling to specialty shops and cooperatives. Unfortunately, small farms, even when banding together, cannot meet the food demands of modern society. Farmers simply cannot grow and harvest that much food on small patches of land dedicated to a variety of different crops. The conventional food system is vastly more efficient, with acre upon endless acre of one crop—corn, wheat, potatoes—protected from bugs by ample pesticides, harvested and packaged in one fell swoop.

By the mid-1970s, in the wake of environmental abominations such as the flammable Cuyahoga River near Cleveland, consumers began demanding organic foods, and organic farmers couldn't meet this demand. It wasn't long before big business saw big profits. So the feudal lords of agriculture—Archer Daniels Midland, Dole, and others—created or acquired organic farms. The U.S. Department of Agriculture helped them along with a lax set of organic standards that allowed the use of sewage sludge, irradiation, and genetic modification under the "organic" label. By 1997, outraged organic farmers (the real ones) organized letter-writing campaigns and passed around petitions in the marketplace to rally the public to protest the new standards. Sewage sludge is now out of the organic standards, as are antibiotics and hormones. A victory? Maybe not. Organic foods can still be processed and rendered into a mash that is as nonnutritious as conventional foods. The unholy trinity of sugar, salt, and fat is completely organic. There are organic TV dinners, after all. And some people buy them because they think "organic" means "healthy."

All the word *organic* means in the food industry is that plant-based food is not grown with synthetic pesticides and that livestock are not fed nonorganic food during the few months to years they are alive. Organic foods can still be polluted in ways that are different from yet just as harmful (or unharmful) as conventional foods. The organic manure adds lead, arsenic, and other potentially toxic metals to plants. Unlike pesticides, metals do not wash off. Also, all food contains dioxin and other pollutants from the atmosphere. The organic product itself may be the epitome of pollution: junk food can be organic provided that most (not all) of the ingredients are organic. White bread and Twinkies—the bane of nutritionists—can be organic. Breakfast cereals can be organic despite their generous portions of processed sugar, salt, and bleached flour. Milk can be organic, even if it is "ultrapasteurized," a process that increases the milk's shelf life and allows it to be shipped across the country. Ultrapasteurized milk is heated beyond the regular pasteurization level to kill bacteria—also destroying vitamins and enzymes. Fruit can be organic even if it is grown under a military dictatorship that chops off the hands of workers caught eating the fruit. Dairy products and eggs can be organic even if the animals that produced the food lived shackled lives. Chances are—unless the label says very specifically other-wise—the cows and chickens were treated in the same brutal, mass-produced manner common to the conventional food business.

Synthetic pesticides can indeed cause cancer, but the risk is very low. The Environmental Protection Agency requires that pesticides carry no higher than a one-in-a-million risk of cancer. (You have about a one-in-a-hundred, or 1 percent, risk of choking on your food; just ask the second President Bush.) Alar, an artificial growth regulator, rocked the world of conventional apple growers in 1989 after the National Research Defense Council published a report entitled "Intolerable Risks: Pesticides in our Children's Foods," alleging that Alar exposure would cause an additional six thou-sand cancers in children. Seeing how childhood cancer is rare and how six thousand was double the yearly incidence, the report was certainly cause for alarm. The CBS news program 60 Minutes ran a hyped-up version of the story, as it was spoon-fed to them by the NRDC. Panic ensued. Schools banned apples. Applesauce and apple juice spoiled on the supermarket shelves. The apple market bottomed out. Many small farmers lost their farms, even though only 15 percent of the apples on the market were sprayed with Alar. The EPA labeled Alar as a potential carcinogen, more for political reasons than for the science. (Only a few questionable ani-mal studies showed that Alar was harmful, and only at *very* high doses.) This was a pyrrhic victory for environmental groups. Alar is gone, voluntarily pulled from the market, but many independent apple growers have disappeared, too. Worse, organic apple grow-ers invested heavy capital, thinking that organic apple sales would rocket. The boom never came; the Alar scare faded quickly, and folks went back to buying conventional apples. Many small organic apple growers also lost their farms. So the scare did nothing but hurt the little guy, the kind of farmer the NRDC likes. And in the end, Alar likely didn't and wouldn't cause cancer.

Wealthier folks are more apt to buy organic foods because they can afford them. The foods offer them piece of mind. Yet, are these organic shoppers making rational health decisions? If they treat their suburban lawns with chemicals to keep them green, then they are exposing their children and neighbors to far higher levels of potentially harmful pesticides and herbicides than that found in nonorganic foods. If they are driving gas-guzzling sport utility vehicles, then they are exposing themselves to carcinogenic, benzene-laden gasoline fumes with each fill-up. They are also adding more than their share of deadly pollutants to the atmosphere.

After thirty years, no study has shown that eaters of organic food are healthier than eaters of conventional food. No centenarian in the United States grew up with organic food. Aveline Kushi, one of the founders of the macrobiotic lifestyle that includes only organic vegetables and grains, developed cervical cancer at age sixty-nine and died nine years later in 2001. This is the diet that many swear by as a cure for cancer, not just a preventative. Kushi, a peace advocate as well, is credited with single-handedly popularizing the "all natural" food movement in America. But as she herself would have attested, there are clearly no guarantees in life.

If not pesticides, what's the real food risk? You are far more likely to die or at least get sick from eating food with harmful bacteria than from eating any food with pesticide. The organic label cannot protect you from this. The major food-borne bugs—salmonella, *E. coli*, listeria, and campylobacter—are found equally on conventional and organic foods. Estimates vary wildly, but even the most conservative seem grim. The U.S. Department of Agriculture estimates that 40 percent of chickens contain harmful bacteria; the Food and Drug Administration puts it at 60 percent. *Consumer Reports* found that 71 percent of chickens that magazine staffers analyzed contained harmful bacteria; and the Minnesota Department of Health found that 88 percent of chickens in that state contain campylobacter. Some amount of *E. coli* is on virtually all chicken sold in the United States. The Centers for Disease Control and Prevention estimate that food-borne bacteria cause 5,000 deaths, 325,000 hospitalizations, and 76 million illnesses (such as diarrhea) each year. Cooking food will kill bacteria, but we do not often cook lettuce and sprouts, the two vegetables most likely to be contaminated.

Hard to imagine that the American food supply can be so contaminated. The problem is that food is mass-produced and rarely locally produced. Remember, organic no longer means local. Food is grown and processed very far away and passes through many hands in its journey from farmland to supermarket. Each of those hands can contaminate the food product. The main source of bacteria is the food processing plant. These factories process epic amounts of food. Take the beef industry. Tons of live cattle are transformed into tons of hamburger at a rate of several heads per minute. This is a bloody, smelly business that tends to attract bacteria. Some bacteria come from manure on the cows or in their guts. Some bacteria come from workers who don't wash their hands after leaving the restroom. Some bacteria come from dirty trucks. Some bacteria grow when meat is not properly refrigerated. Some bacteria come from the busy butcher block in supermarket distribution hubs. Some bacteria flourish simply because the meat has traveled for thousands of miles over the course of weeks. There is often no difference between organic and conventional meat in this regard. Once, you could trust your local butcher, but the local butcher has been put out of business.

Some advocate irradiating meat at the processing plants to kill bacteria. This is not a complete solution, for meat can easily be recontaminated en route to your frying pan. The best defense against food-borne bacteria is to wash food thoroughly and to cook it well. The second-best defense is to buy local meat and produce, which is fresher and has passed through fewer hands. Buying organic won't spare you from food-borne bacteria. In the defense of organic fruits and vegetables, they are not *more* likely than nonorganic food to contain bacteria because they are fertilized with manure. Claims to the contrary are as yet unfounded.

The argument against conventional farming is that tons upon tons of synthetic pesticides and fertilizers poured into the soil year after year can sicken beneficial insects and microorganisms. The pesticide eventually enters into groundwater supplies or washes directly into waterways. This is true, and it may be a problem. Organic farming is kinder, although it requires excessive tilling to control for weeds. This depletes oxygen, nitrogen, and essential elements from the soil. The compromise, since we have chosen to make organic farms large and cannot weed them by hand, is to kill the weeds with propane torches, which themselves emit toxic exhaust. Overall, organic farming seems more sustainable than conventional farming, but no one is sure about how *un*sustainable conventional farming is. Also, no one is sure if organic farming can feed the world during a time of famine or insect plague. When was the last time you heard about a locust plague in the United States? Locusts devoured the entire Dakota harvest over the course of only a few days in 1867 and periodically in the decades that followed before the era of pesticides.

There are two advantages to buying organic. Organic wholesalers and grocers often enter the business because they care about food. Thus, their markets tend to offer a wider range of fresh vegetables and healthier foods. Conventional supermarkets, by and large, have a paltry selection of vegetables and take less care in storing and displaying them. The food at an organic shop is often healthier by virtue of its freshness, cleanliness, and diversity. And like bottled water versus tap, it's a matter of taste. Supporting organic supermarkets supports the notion of respecting and caring for food. Buying organic also supports the folks harvesting the organic food who, unlike those on conventional farms, are spared the serious health risks of acute pesticide poisoning.

Perhaps those enamored of that original 1970 definition of organic—good to the land, good to the animals, good to the workers—might be persuaded to support local farming as opposed to strictly organic farms. After all, the local farmer isn't evil just because he uses a pesticide or herbicide to control bugs or weeds. Local, small farms plant those vegetables best suited for the local environment. Without support, these farmers will be forced to sell their land to real estate developers to make a bland housing development or a shopping mall—and all our food will be grown, albeit organically, in the California central valley, a natural desert artificially flooded by diverting the waters of once-mighty rivers.

25

ξ ξ ξ ξ ξ ξ ξ

Water, Water Everywhere:
Bottled Water vs. Tap Water

Is bottled water is healthier than tap water? Not necessarily. Much of the bottled water sold today is actually municipal tap water, filtered in a factory to varying degrees, and given a fancy name like "Glacier Springs" or "Aqua Expensivo." Most of the time, tap water is safe (depending on what city you live in) and healthier when you consider that fluoridated water prevents tooth decay and therefore cancers associated with oral infections.

If you visit Mexico, the saying goes, don't drink the water. Americans apparently take this message to heart in their own country. Sales of bottled water in the United States reached over $5 billion in 1999 (and $35 billion worldwide). According to the National Resources Defense Council, which conducted a four-year study on drinking water, published in 1999, the majority of Americans surveyed drink bottled water because they view it as healthier than tap water. Folks without much money were just as likely to buy bottled water as the rich folks, for they were even more skeptical of their poor neighborhood's water supply. Bottled water costs 250 to 10,000 times more money than tap water, which is supplied in ample amounts in most U.S. homes for a fraction of a cent per gallon. Is paying premium for fancy water worth it? In terms of taste, yes. In terms of health, no.

The comedian W. C. Fields, a notorious boozer, wouldn't drink water because "fish piss in it." No drinking water is entirely pure.

Depending on the region, there are varying amounts of minerals and metals in all water. Tap water in the United States, regulated by the Environmental Protection Agency, is surface water drawn from lakes and reservoirs or, in rural areas, groundwater. The water is treated for bacteria in processing plants (usually with chlorine) and must meet certain purity levels for harmful pollutants such as arsenic or lead. Pollutants are measured in terms of one part per million, billion, or trillion molecules of water. Not every pollutant is removed; it's too expensive and, beyond a certain level of purity, doesn't make the water any healthier. Why screen all pollutants to the one-in-a-trillion level at an increased cost of millions of dollars—money that could go to, say, schools or police?

Bottled water is regulated as a food by the Food and Drug Administration with a different set of criteria. As a result, some bottled waters have higher levels of bacteria and metals than the EPA allows for tap water. Remember the $130-million Perrier recall in 1990 for exceeding benzene limits? (Neither does anyone else; good PR.) But before we condemn the folly of buying bottled water for health reasons, let's first examine the possible negative side of tap water.

Big cities usually do a good job of providing safe drinking water, but there have been some recent goof-ups. Washington, D.C., the nation's capital, has had notable municipal water problems. In 1996 the U.S. Army Corps of Engineers found high bacteria counts throughout the entire D.C. water system. The city responded by upping the chlorine level to kill the bacteria. Chlorine is an effective cleaner, but it contributes to tap water's foul taste, and the chemical can combine with other molecules to create cancer-causing agents, as is explained later in this chapter. Excess bacteria creeps back into the D.C. water supply from time to time.

Healthy folks can handle the little bugs, with names such as *Cryptosporidium* and *Giardia*. Children, the elderly, and those with compromised immune systems do not fare so well. For these groups, the bacteria can cause severe abdominal pain, diarrhea, and death. A *Cryptosporidium* outbreak in Milwaukee, Wisconsin, in 1993 killed over 100 people and sickened an estimated 400,000. The bacteria likely came from manure from cows grazing near the streams that feed Milwaukee's water supply, good ol' Lake Michigan. The system could filter industrial pollutants in the lake (the kind that killed the fish) yet missed *Cryptosporidium*. A true tragedy.

Tap water can indeed have a foul taste from chlorine. Yet, the majority of Americans cannot tell the difference between tap water and bottled water; taste tests show this time and again. We buy the pretty bottle; and the prettier it is, the better the water tastes to us. It's largely mental. Most municipalities see chlorine as a very positive trade-off. The stuff kills bacteria, which otherwise cause instant problems—serious illness and death for many people during outbreaks. This is how public health works. You address a serious problem—in this case, chronic water contamination for millions of people—with a blanket solution: chlorine. Bottled water isn't a solution. We cannot give everyone bottled water. The advantages of treating water with chlorine grossly outweigh any minuscule risk that chlorine will cause cancer. Chlorine's byproducts, namely trihalomethanes such as chloroform, are only *probable* human cancer-causing chemicals when present in high levels—causing *possible* long-term cancers in a few people. Trihalomethanes form when chlorine reacts with organic molecules in water. How much is too much trihalomethane? California, for one, sets a safety limit at ten parts per billion (a number exceeded by some bottled waters, by the way). Most exposure to trihalomethanes, oddly enough, comes not from drinking cold water but from showering in hot water. You breathe it in. A few municipalities are searching for alternatives to the very effective chlorine treatment, such as ozone treatment, which, believe it or not, is safe.

Swimming pools have far higher amounts of chlorine than drinking water has, yet no one seems to care. If you can smell the chlorine, then you are breathing it in, ingesting it. One must assume that swimmers lounging around the pool with Perrier and Aqua Expensivo are drinking them for the taste, for their pool exposure to chlorine dwarfs any dose they would get from tap water. If chlorine in drinking water is causing cancer, it must be minimal, for no health study has found statistically significant risks. And a lot of tap water is being drunk. How deadly is *no* chlorine? The World Health Organization estimates that 25,000 children die daily worldwide from illness caused by contaminated water. Adequate chlorination and better hygiene would prevent this. In Peru, a breakdown in water quality control and lack of chlorination led to a ten-year cholera epidemic that spread to neighboring countries and killed over 15,000 people during the 1990s.

The metal lead is often cited as being a problem in tap water. Water may leave the processing plant clean and travel through most of the city, but it gets contaminated with lead from old pipes in old homes. Some new homes have pipes that are lead-free but do contain lead soldering. Lead poisoning can cause learning disabilities in children, among other things. Remember, though, that you and your parents grew up in a world of lead paint and lead gasoline. To imply that lead in drinking water will cause learning disabilities is to suggest that most Americans have been exposed to amounts of lead that would have made them imbeciles. (Hmmm.) Anyway, lead leaches from pipes and soldering, particularly in hot water, so you can cut your lead exposure significantly by letting the water run cold before drinking or cooking with it. Out in the countryside, folks get their water from wells, and regulators aren't around to test for bacteria, metals, pesticides, or even gasoline. You're on your own. Fortunately, you can buy home-testing kits and filtering systems.

Arsenic dominated the news in 2001 when the EPA, led by director Christine Whitman, considered abandoning the plan to lower arsenic levels in drinking water to 10 parts per billion (ppb). For decades, the arsenic level was capped at 50 ppb, though there was growing concern that this was leading to increases in bladder and lung cancers. The National Academy of Sciences released a report in September 2001 stating that even the 10-ppb level might not be low enough. Most cities have arsenic levels lower than this; the problem areas are in rural America, where arsenic is a by-product of mining. Some 13 million Americans are affected. So arsenic, although an issue for some, is not a reason for most Americans to turn to bottled water.

Actually, there is no guarantee that bottled water will be free of lead, arsenic, or bacteria. In fact, bottled water rules permit a certain level of *E. coli* or fecal coliform contamination. Tap water rules have zero tolerance for these pollutants. The 1999 report from the National Resources Defense Council, which tested 103 brands of bottled water, found that most were of high quality; yet one-third contained levels of contamination—synthetic organic chemicals, bacteria, and arsenic—that exceeded allowable limits under either state standards or bottled water industry guidelines. Other tests have also come up with this number of about one-fourth to one-third contamination, such as a survey in 2000 by the School of Dentistry at Case Western Reserve University in Cleveland, Ohio. Researchers there found that 15 out of 57 bottles tested had 10 to 1,000 times the bacterial levels of Cleveland water plants. Bottled water companies may say their products contain no chlorine or harmful elements, but they are wrong.

Just what is bottled water? The International Bottled Water Association estimates that 25 to 40 percent of bottled water sold in the United States comes from municipal supplies—the same water that comes out of your tap, only filtered differently. Usually the label says "filtered water" or "purified water." A popular brand is Aquafina by Pepsico. Coca-Cola also packages filtered water. Usually these products have a picture of a pretty mountain on the label instead of a water-processing factory. Some exaggerate just a little too much: Alasika called itself "Alaska Premium Glacier Drinking Water: Pure Glacier Water From the Last Unpolluted Frontier, Bacteria Free." The FDA made the company change the label upon learning that the water came from a public supply.

Spring water is bottled water that comes from springs, such as Poland Spring. Well water is water that comes from aquifers, underground pockets of water. Distilled water is pure H_2O and has no nutritional value. Mineral water is spring water or well water that contains at least 250 parts per million of dissolved minerals. Sparkling water is weird. This is water with natural carbon dioxide (the stuff that makes the bubbles), which often needs to be removed during the purification process and added again so that the water contains the same amount of carbon dioxide that it had at its source. These types of bottled water are safer than tap water, if at all, solely by virtue of their having less chlorine, but again, there is no guarantee. Perrier mineral water comes from a variety of sources beyond France, such as Texas and New Jersey. Somehow, somewhere, in 1990, unacceptable levels of benzene, a known cancer-causing chemical, made their way into the stylish green bottles. The benzene level was far from deadly or even cancerous. You would have had to drink a couple hundred bottles of Perrier a day to get a level that would significantly increase your lifetime risk of getting cancer; and by that time, at $2 a bottle, you would have died of poverty. The Perrier recall, however, did make Americans question the integrity of bottled water . . . at least for a few months.

Times have changed. The bottled-water industry has seized upon the fear that we have of inadequate, aging, and dangerous municipal water supplies. The fear comes naturally with occasional reports of lead or bacteria in these water supplies. The reports are not overstated; it is the interpretation of the reports that leads us to believe that all tap water everywhere and anytime is dangerous. Bottled water is a nice and even useful alternative to tap water, particularly during times of bacteria breakout. Bacteria can overload a water-processing plant in times of floods, when raw sewage mixes with the water reserves. That's a good time to reach for a $1.50 bottle of water. Another good time is when you enjoy the taste of bottled water. One cannot argue about taste; this is subjective, unlike health consequences. (Interestingly, the city of Houston plans to sell its tap water in supermarkets in bottles, straight, without additional filtering. We'll see how consumers respond.)

One thing that bottled water does not have is fluoride, which is added to tap water solely to prevent tooth decay. Sounds like a crazy idea, but it works very well. The few communities that do not have fluoride in their water have significantly higher rates of tooth decay. Many health officials consider fluoridation of the water supply one of the greatest health achievements of the twentieth century, right up there with penicillin and vaccines. Having healthy teeth is far from being merely a cosmetic goal or a snub to the British, some of whom look like they brush their teeth with a Zagnut Bar. Tooth and gum decay can lead to ulcers, certain types of cancer, and a general decline in the immune system. Kids, especially, need fluoride, and if they don't get it from water, they'd better be getting it from fluoride tablets and toothpaste.

The actual cost of the water placed in fancy bottles is at most a few pennies. The Defense Council report found that typically 90 percent or more of the consumer cost goes to bottling, packaging, shipping, marketing, retailing, and, naturally, profit.

PART V

The Return of the Witch Doctor

Medicine is the most distinguished of the arts, but through the ignorance of those who practice it, and of those who casually judge such practitioners, it is now of all the arts by far the least esteemed.

—Hippocrates (460–400 B.C.E.)

Much of the time, complementary and alternative medicine isn't complementary, alternative, or medicine. Other than that, the name's right on the money. Clearly, as a society we feel that some ancient traditions are passé. We are far too sophisticated to gargle with goat urine or submit to bloodletting. Yet why do we yearn for other ancient cures and customs born of the same logic, from an era when most people died young from diseases we've since licked?

Don't be fooled by the lure of alternative medicine, boasting of relaxation, exercise, and a healthy diet. This is not alternative medicine. This is commonsense mainstream medicine. Alternative medicine simply pulls you in with this. Beneath the surface, alternative medicine is weird and potentially deadly stuff. If it worked, it would not longer be called "alternative."

My yoga instructor had a very serious reaction to a bee sting. Here's a guy who is a vegetarian and normally reaches for natural herbal cures, eschewing conventional medicine. Yet when faced with a pulse rate of over 200 and an inability to breathe in the hospital emergency room, he decided to forgo tea tree oil and all the other nonsensical "alternative" cures and readily agreed to an epinephrine injection. This saved his life. He now carries the epinephrine and an antihistamine around with him today, and he cheerfully relates the hilarious story of how he quickly abandoned alternative medicine in the face of death. This is the heart of the issue. Traditional cures were once the only cures. By and large, most simply didn't work. That's why we abandoned them. There's no drug-company conspiracy here. Just because a cure is ancient or exotic (likely from a country with a low life-expectancy rate) doesn't make it useful.

Even the man on the street in China is no fool. He understands that Viagra works better than ground-up rhino horns for erectile dysfunction. Viagra may save the rhino from extinction. Likewise, the governments of China and India—the birthplaces of so many alternative therapies—are trying to do away with traditional cures in an effort to increase life expectancy. They are fighting for this in the United Nations every day. Give us vaccines and effective medicines, they demand, not more sandalwood-scented candles. Today, the main consumers of alternative medicine are Westerners, several generations removed from remembering how hard life can be.

26

ቜ ቜ ቜ ቜ ቜ ቜ ቜ

The Delusion of Dilution: Homeopathy x 50

Homeopathy uses two basic principles: like cures like; and dilution is the best solution. What strikes me about homeopathy is how popular it is among wealthy, apparently well-educated people. I visited one large, bustling health-food and homeopathy store occupying a desirable plot of real estate in Harvard Square in Cambridge, home to one of the most respected universities in the world. (Harvard University is nearby, too.) I wondered whether any of the patrons—no doubt very scholarly students, professors, and white-collar professionals—understood homeopathic principles.

Like cures like. Baby rash is cured with a diluted treatment of poison ivy. Fevers are treated with herbs and roots that make you warm. Stiffness is treated with snake venom. The father of homeopathy, Samuel Hahnemann of Germany, spent the close of the eighteenth century determining which natural substances produced which symptoms and then recorded them as possible cures for people with these symptoms. What got Hahnemann cooking on this "like cures like" notion was the fact that quinine, a malaria drug, induced malarialike symptoms when taken straight by healthy people. And a pseudoscience was born.

There is nothing intrinsically illogical about "like cures like," other than the fact that it doesn't work. Homeopathy crosses the line of reality, though, when it comes to dilution. Hahnemann called it the "law of infinitesimals." I call it delusional dilution. The homeopathy cures worked best, Hahnemann found, when they were diluted. Much of this medicine, after all, is very toxic, so minimizing the dose would lead to fewer side effects. Hahnemann would dilute the medicines so much that they essentially became water. Yet water, it seemed, was better than the other ridiculous cures around at the time: bloodletting, arsenic, mercury, and other traditional medicines that didn't work. Homeopathy grew in popularity because it did no harm and, with the placebo effect and the body's own ability to heal itself, appeared sometimes to work. After 100 years, though, homeopathy was just about dead once people realized that (a) it wasn't really curing anything consistently, and (b) they were paying for sugar water.

In the 1930s, inexplicably, homeopathy made a comeback. U.S. senator Royal Copeland, a homeopath, wrote a provision that exempted homeopathy from the Food, Drug and Cosmetic Act of 1938, so the cures didn't need to be proven safe. No big deal; they're water, after all. Unfortunately, the Copeland provision lived through the 1964 amendment to the 1938 law, which stated that therapies must be proven effective. Homeopathy is now mainstream medicine in Germany and is catching on in the United States among people who don't get enough sugar water in their soft drinks.

You want to talk about dilution? Look on any bottle of a homeopathy cure, and you'll see the dilution level. A few say 30x. In this case, x stands for ten. A 30x solution starts at 1 part medicine per 10 parts sugar water or alcohol. This is mixed. One part comes out and is mixed again with 10 parts liquid. One part comes out again. This is repeated 30 times. The remaining solution is, according to this formula, one part medicine per 10^{30} parts sugar water. For annoying redundancy, I'll write that out: 1 per 1,000,000,000,000,000,000,000,000,000,000. This is far beyond the dilution capacity of the vial of homeopathy medicine. The physicist Robert Park, the sardonic author of *Voodoo Science*, estimates you would have to drink 7,874 gallons of the solution to get one *molecule* of the medicine.

It gets worse. Most homeopathic medicine in every health-food store that I have been to has a dilution set at 30c. This is one part medicine to 100 parts sugar water, mixed and diluted in this manner 30 times. That works out to be one part medicine per 100^{30}, or 10^{60} parts sugar water. I'll spare you the zeros. But turning again to Robert Park's math, the entire universe contains about 10^{80} atoms. With this dilution, you'd have to drink in the entire solar system just to get one molecule of medicine in your body. Other homeopathy solutions are 100c, far beyond the universe's ability to accommodate. Anything higher than 24x, actually, is implausible.

Early followers of homeopathy did dilute the medicine over and over again with no concept of dilution capacity: that is, one remaining molecule of medicine per a given amount of water. Today, with the help of Avogadro's number, we know how to calculate the number of molecules in a given substance. So modern makers of homeopathy medicine understand that they have exceeded dilution capacity. They just don't care. Yes, the makers of homeopathy cures will admit that there is no medicine in the medicine. Their reasoning is that the water somehow remembers the shape of the medicine molecules. This memory is maintained whether the homeopathy cure is in liquid or pill form, and the memory lingers as the solution is broken down by your body.

It gets worse. A leading proponent of homeopathy, Jacques Benveniste of France, claims that the medicine-induced shape of the water can be captured electronically, stored digitally, and sent across the Internet to be downloaded into other vials of sugar water. None of this would be so bad—after all, no one's dying from sugar water—if it weren't for the fact that Benveniste caught the sympathetic ear of Wayne Jonas, the original director of what was then called the Office of Alternative Medicine at the National Institutes of Health. (They have since added the word "Complementary" and became a Center, minus Jonas, who has "moved on.")

The fact that an NIH director would seriously entertain such a notion is frightening, for Jonas controlled the purse strings for research into alternative medicines for nearly four years. This water-memory logic would imply that all water is therapeutic. Dioxin causes cancer. Removing dioxin from the public drinking water supply leaves the memory of dioxin. Like cures like in highly diluted concentrations, right? So water stripped of dioxin is a cancer cure. If water can retain a memory of a given medicine, then this is totally new physics. Homeopathy believers now point to the world of subatomic particles and bizarre quantum forces and phenomena that even physicists cannot fully comprehend. Maybe the memory is trapped in there, they say. This is a common ploy among alternative medicine shamans. They introduce intelligent-sounding modern scientific theory to explain their magic. There's really no logical end to this argument: quantum fluctuations cause subatomic particles to fleet in and out of existence, therefore I can walk through walls. Jonas himself has written that chaos theory might account for the homeopathy effect.

If homeopathy worked, then advocates might have an argument. After all, they admit that no one can explain *how* homeopathy works, just as the mechanism of conventional cures cannot be explained. True. Yet, demonstrating homeopathy's therapeutic effect, which should be easy, has been a tough go. Jonas summed it up wonderfully in his 1996 book with Jennifer Jacobs, *Healing with Homeopathy: The Complete Guide,* a real gem: "Until recently, there has been little research in homeopathy, either in the laboratory or with patients. Physicians using it were busy trying to help patients and had little interest in research." Such a noble breed, obviously overworked with all that diluting.

In this book, Jonas states that the majority of homeopathy studies show homeopathic medicine to be more effective than a placebo. Results are so narrow and tests are so small, though, that this can be explained by chance. Researchers are merely measuring which is more effective, placebo number 1 or placebo number 2. The authors are also generous with their definition of "majority." Negative homeopathy studies are underreported, and positive studies are questionable, including coauthor Jacob's study of homeopathy treatment of chronic diarrhea in Nicaragua highlighted in the book. Published in 1994 in *Pediatrics,* the study has since been devalued for having unreliable diagnostic schemes ("cure" meant fewer loose bowel movements, a subjective determination) and no significance, because time and adequate fluid intake will improve diarrhea.

Today, meta-analysis reports, which compare all published homeopathy studies, have the flavor of this 2000 report by Cucherat et al. in the *European Journal of Clinical Pharmacology:* "There is some evidence that homeopathic treatments are more effective than placebo; however, the strength of this evidence is low because of the low methodological quality of the trials. Studies of high methodological quality were more likely to be negative than the lower quality studies." In other words, the better the study, the more it shows that homeopathy is merely a placebo. Animal studies don't support homeopathy, either, because animals aren't smart enough to be duped by a sugar water. Why are we struggling to show that homeopathy is as good as a placebo, anyway? These homeopathy cures cannot compare in the slightest degree with simple, safe, chemical compounds that can bind loose stools or stop a runny nose. In the realm of allergy, or sore-eye relief, homeopathy is innocent enough. Yet treating serious diseases such as measles with homeopathy is downright unconscionable

Another popular homeopathy defense—aside from "The world is mysterious," or "Water memory is beyond our humble comprehension"—is that highly diluted substances can have an effect on the body. No kidding; that's why we screen chemical pollutants in the drinking water to the one-part-per-billion/trillion level. Highly diluted, though, is different from implausibly diluted, and that's homeopathy. All of these illogical assumptions and shoddy investigations into something that it should be simple to prove effective are, perhaps, taking their toll on Jonas. In a June 2001 report in the *International Journal of Epidemiology,* Jonas and his colleagues wrote: "Trials of complementary therapies often have relevant methodological weaknesses. The type of weaknesses varies considerably across interventions." This is the full conclusion in their abstract of a report investigating how homeopathy studies are performed. At least he's honest.

Homeopathy is fun because you get to play chemist. You take the medicine and dilute it over and over again yourself. Shake, shake, shake. Dilute. Shake, shake, shake. Dilute. This in itself adds to the placebo effect, because you are creating the cure, just like a shaman. The odd thing is that homeopathy is not a snake-oil cure duping the unschooled farmer, like traveling medicine shows in the past. Rather, folks of all educational levels and income brackets are enticed by homeopathy's charms. Shake, shake, shake.

27

ʥ ʥ ʥ ʥ ʥ ʥ

Magnetic Charm:
Magnets and Your Health

Magnet therapy is charming America, just as it captured France and Austria over two hundred years ago before this obvious fraud was uncovered. It is ironic that the lure of magnets is so strong, because therapeutic magnets themselves are rather weak. Therapeutic magnets don't have enough strength to penetrate their holders, let alone your skin. The magnets have no magnetic effect beneath your skin, which is why the physicist Robert Park jokingly calls them "homeopathic." I am fascinated by another similarity with homeopathy: magnets ain't cheap. The wealthiest of Americans are the first to surrender their money for them. I was once annoyed about paying $5 for a souvenir refrigerator magnet at Fisherman's Wharf in San Francisco. Therapeutic magnets can go for up to $100.

Magnet therapy is based on one simple fallacy: iron-rich blood is attracted to magnets, and this improves circulation. The iron in your blood is locked up in hemoglobin molecules, which are actually slightly repelled by magnets. Sure, you can wear a magnet someplace on your body all day long and notice redness in that area. This is because you have a block of metal strapped to your arm. The magnets aren't causing blood to come to the surface; the heavy weight and the stress of carrying a chunk of metal are the cause.

If blood cells were attracted to magnets, then magnetic resonance imaging (MRI) scans would kill you instantly. An MRI scan

164

is a technique used to image soft tissue inside the body. These machines create a strong magnetic force. You have no doubt seen these futuristic-looking MRI devices—big white things in which a patient lies on his or her back and slides into a central tunnel. You may have also heard the story of that poor child who was killed during an MRI scan in 2001. Somehow, a fire extinguisher was left unsecured and exposed to the MRI's magnetic field. This twenty-pound chunk of metal went flying across the room and smashed the head of a young child undergoing an MRI scan. If magnets affect blood flow, then the same force capable of turning a fire extinguisher into a missile would cause the delicate veins in your body to explode.

Even if magnets could affect blood flow, which they don't, the magnets you buy at New Age shops aren't strong enough to penetrate the skin. This is easy enough to test. Take one of those magnets and see if it can support a shirt on a refrigerator. Chances are it won't, and your skin is thicker than a shirt. See if the magnet can pick up a paper clip while it's in its Velcro™ strap. Chances are it won't, which should leave you wondering what possible effect the magnet could be having on the body. Some "magnet therapists" claim that the magnets improve the flow of *chi,* a Chinese concept of vital energy flow. The *chi* gets the blood moving. The therapists are just grasping at straws at this point. If magnets affected *chi,* then we would at long last have a method to measure *chi,* an abstract concept. The serious Chinese *chi* therapists I've interviewed over the years, called *qigong* masters, all think Western magnetic therapists are wacky. They keep their distance.

Nonetheless, the sports world is hooked on magnets. Magnetic shoes sell at golf pro shops for well over $100. (Just what you want when lightning strikes.) The idea here is that the magnets draw blood to the feet and improve circulation, relieving fatigue on that tough six-mile drive in a golf cart around the links. Dan Marino, a former quarterback for the Miami Dolphins, claims that magnets helped heal his fractured ankle. Sports trainers have strapped magnets to just about every part of athletes' bodies to reduce muscle fatigue, heal fractures, or otherwise speed recovery. Estimates of U.S. annual sales of therapeutic magnets range from

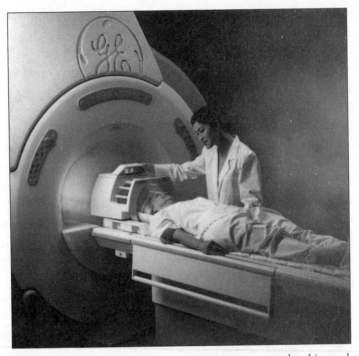

The force from therapeutic magnets cannot penetrate the skin, and even if it did, it would have no effect on blood circulation. Blood isn't even attracted to the magnets from MRI machines, which are millions of times more powerful than therapeutic magnets.
Courtesy of GE Medical Systems

$200 million to $500 million, up from only a few million dollars per year in 1990.

Pulsed electromagnetic fields in a hospital setting can effectively treat slow-healing fractures. These pulses, though, are from complicated electrical devices, not refrigerator magnets on Velcro straps. The magnetic-therapy industry can cite only one positive strap-on study, a small pilot study from Baylor College of Medicine in Texas that showed that people with chronic knee pain were more likely to feel slightly better after wearing a magnet compared to the placebo group. The authors of the study themselves said this was just a pilot study to see if the magnet issue was worth investigating in a larger, better-defined study. Other studies did indeed follow this, and they found that strap-on magnets had no effect on healing.

Whereas homeopathy enjoys mild popularity among certain medical professionals (the same five researchers, it seems), magnet therapy and its sister therapy, crystal healing, are beyond the fringe. No respectable researcher will come out in support of magnet therapy or crystal healing. In the book *The Practical Guide to Magnetic Therapy*, author Peter Rose states that "the lost kingdom of Atlantis used crystal power to provide most of its energy needs." Apparently not enough to keep the city afloat. The Federal Trade Commission finally cracked down on some of these health claims in 1999. The Texas-based Magnetic Therapeutic Technologies was barred from claiming that its products were effective in treating cancer, HIV, diabetes, arthritis, and several dozen other ailments. The New York–based Pain Stops Here! was barred from claiming its magnets could help treat infections, organ and circulatory diseases, muscle and joint problems, and dysentery.

The history of magnet therapy is as old as the discovery of magnets. Admittedly, magnetism seems like a bizarre force. Magnets can levitate trains, after all. Surely, you would think, they can have a powerful effect on the body. They may, but nothing sold in stores comes close to having the type or level of magnetic force capable of altering anything in your body. Franz Mesmer of Vienna was one of the first quacks to fool the public with magnetic healing. (The word "mesmerize" stems from his name.) No less a scientific authority on electromagnetism than Benjamin Franklin hounded Mesmer during his Paris demonstrations in the late 1770s. Franklin helped form a commission to investigate Mesmer, found him to be a fraud, and ran him out of town. Today, magnetic healing devices are sold at Brookstone, a dealer of fun and exotic objects with a sci-fi feel, as well as at other businesses that pride themselves on promoting science and technology. Where's old Ben when you need him?

28

ᵇ ᵇ ᵇ ᵇ ᵇ ᵇ ᵇ

Reversal of Fortune:
The Viability of Ayurveda

Thirty-odd years after the Summer of Love, America is once again enamored of the guru. Ayurveda, a modern incarnation of a nearly forgotten ancient Hindu healing art, is attracting a growing number of largely wealthy Westerners who spend thousands of dollars on herbal concoctions and seminars offering advice on how to grow younger. At best, Ayurveda is a healthy lifestyle that promotes a vegetarian diet, yoga, and relaxation. At worst, Ayurveda is a multimillion-dollar business of sham cures based on astrology, gem healing, psychic healing, mantras, and the faulty science of bodily humors, spun through either fraud or naiveté.

Maharishi Mahesh Yogi, the famed guru to the Beatles for a few months before they quickly grew disenchanted, ignited the modern Ayurvedic movement in 1980 as an offshoot of Transcendental Meditation. There are many flavors today. Vasant Lad, the director of the Ayurvedic Institute in Albuquerque, New Mexico, promotes a brand of Ayurveda that emphasizes herbs, oils, incense, and horoscopes. Deepak Chopra, author of such best sellers as *Ageless Body, Timeless Mind: The Quantum Alternative to Growing Old*, stresses positive thinking, a practice he calls mind-body healing. As with other proponents of alternative medicine, many leading teachers of Ayurveda sell the very products they deem as crucial for good health and disease prevention. For example, the Maharishi, far from being an impoverished guru, maintains a com-

168

mercial web site at http://www.maharishi.co.uk, where one can purchase organic foods, oils, aphrodisiacs, books, and much more. Chopra, a Western-trained physician and former chief of staff at New England Memorial Hospital, was earning $25,000 per lecture by the end of the 1990s. Many lectures and subsequent videos sold through web sites and at New Age stores are particularly targeted at the upper classes, claiming as they do that Ayurveda can even improve one's golf game. Ayurveda's teachers offer countless testimonials from patients who have cured themselves of the incurable.

So what is Ayurveda all about and how can it help you sink a twenty-foot putt? Ayurveda's core concept is that the body is defined by three "irreducible physiological principles," called vata, pitta, and kapha. As with the European notion of the four humors or the Far Eastern philosophy of yin and yang, you have to keep the three Ayurveda forces balanced for harmony (which means nothing, medically) and good health. Unbalanced vata, for example, can lead to constipation, arthritis, and lots of other seemingly unrelated things. An Ayurvedic practitioner takes your pulse to determine your levels of vata, pitta, and kapha, and then prescribes the right diet, herbal regimen, and incantation to get things back to normal. Colds are not caused by viruses per se; they are caused by an imbalance. Cancers are not caused by pollutants per se; they are caused by an imbalance. This is why placing the forces in balance will heal the body. As Virender Sodhi, the director of the American School of Ayurvedic Sciences in Washington State, puts it: "Disease is the result of a disruption of the spontaneous flow of nature's intelligence within our physiology. When we violate nature's law and cannot adequately rid ourselves of the result of this disruption, then we have disease." The alignment of the planets is important as well. According to Vasant Lad, "Each planet is related to a specific body tissue. Mars, the red planet, is related to blood and the liver." Venus, you may have guessed, is tied to impotence. So much for germ theory.

Buying into the Ayurvedic belief as promulgated by these experts allows one to journey back to a time when humans did not know what caused disease. You are journeying back to the era of the shaman. Despite their claims, Ayurvedic healers cannot diagnose illnesses such as diabetes, ulcers, or cirrhosis by taking one's pulse. Practitioners detect up to fifteen different types of pulses from the wrist that they say correspond to the function of six different organs. But these traditional healers have yet to make correct diagnoses in a scientifically controlled setting. There have been tests, such as those performed by the Committee for the Scientific Investigation of Claims of the Paranormal during its visit to China in 1998. The healers performed no better than a tossed coin in guessing which of two symptoms a patient had. They are no better at healing than a mystic healer from ages past. It seems that the diagnosis is 100 percent correct only when there are no legitimate diagnostic tools available to prove otherwise. (The reason the healers consistently founder under scrutiny, as any serious Ayurvedic practitioner will tell you, is that the testing process interferes with their ability to connect mentally with the patient.) Even if the diagnosis is correct, restoring balance, whatever that means, is not how diseases are remedied. And even if getting back in balance is a good thing, you are regaining that balance through a combination of questionable herbs and aromas tinged with astrology.

Ayurvedic healing presents a challenge to all that modern medicine accomplished in the twentieth century. Seeking Ayurvedic treatment for yourself might not be so bad, but we cannot play games with children. Ayurveda manuals offer cures for bacterial and viral infections, such as mumps and measles. Children, including those in India, have died through the ages from such infections, all easily treated today with modern medicine. Ayurveda is weird and potentially dangerous stuff once you scratch the surface. With just a quick glance one learns that the Ayurveda lifestyle emphasizes moderation. Don't eat too much. This is innocuous and even good advice. Ayurveda also promotes positive thinking. This is nice, too. Admittedly, a positive attitude can help sick people feel better and manage their treatment more easily. No study, however, has ever shown positive thinking to prevent or cure disease, and depressed people do not have a higher rate of cancer. Now, dig a little deeper and one will find that Ayurveda promotes aromatherapy and herbal remedies. There might be medicinal value to some herbs, but most remain untested. Of those tested, some seem to

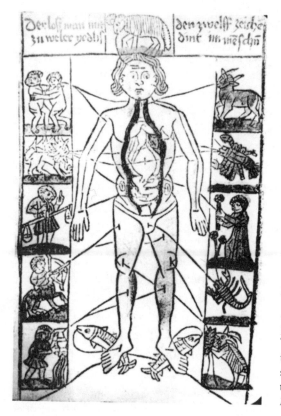

The healing science of Ayurveda connects body parts to astrological signs. This idea was wrong in the Middle Ages, and, not surprisingly, it is still wrong today. *Courtesy of the National Library of Medicine*

have no effect on health, while others, such as kava kava, can be deadly. Still other herbal Ayurvedic remedies are a proprietary mix of herbs and fungi, the exact contents of which remain a trade secret. Dig deeper yet—that is, pick up a book of Ayurvedic cures—and one will discover an unpleasant reality. Some remedies call for animal urine and feces. And why not? This is an ancient practice perfected long before the concept of germs. The standard Ayurvedic remedy for preventing and reversing cataracts is to brush your teeth and scrape your tongue, spit into a cup of water, and then wash your eyes with this mixture. As reported in the *Journal of the American Medical Association,* Ayurvedic treatments can cause hallucinations, anxiety, depression, insomnia, and gastrointestinal problems. Safer treatments include gemstones and prayers, either in the form of a mantra or a ceremony called a *yagya,* but they cost thousands of dollars. Upon a thorough investigation, knee-deep in Ayurveda muck, one will learn that many Ayurveda proponents, such as Dr. James Gordon of Georgetown University, chairman of the White House Commission on Complementary and Alternative Medicine, were either followers or supporters of the late Bhagwan Shri Rajneesh, the cult guru and multi-millionaire who was deported from the United States after his followers deliberately contaminated food at local restaurants with fecal bacteria in an attempt to sicken residents of Antelope, Oregon, and keep them from voting to oust the cult in a local election.

An Ayurveda article in a slick New Age health magazine might not go into such detail. On the contrary, Ayurveda might sound quite exciting and sophisticated, as it incorporates, for example, the concept of quantum physics to explain how meditation, positive thoughts, and Ayurvedic herbs can cure disease. There is, of course, no true connection between quantum physics and Ayurveda. Quantum is a word people use to sound smart, sort of like the stuff you hear in a bad science-fiction flick: "Raise up the alpha-proton shield; the planet is radiating strong gamma forces." The gist of quantum healing in the world of Ayurveda is that the body comprises countless atoms and therefore has an infinite energy supply. (Think nuclear fusion.) Organs, too, are made of atoms vibrating in a specific way. Disease results when the atoms vibrate out of sync, which is manifested as an imbalance of vata, pitta, and kapha. The mind or certain herbs can get these atoms to vibrate correctly. The mind can also reverse the aging process by releasing atomic energy. As any physicist will tell you, this is all nonsense. So popular is the unfounded notion that the mind can cure the body of disease that even the magician and spoon bender Uri Geller has his own health book, *Mind Medicine.* The notable alternative medicine advocate Andrew Weil wrote the foreword, which has put him at odds with many in the health community who have viewed him over the years as a bastion of reason in a field unfortunately tinged with quackery.

Ayurvedic proponents will argue that no one has ever died as a direct result of Ayurvedic treatment, while many have died from conventional drugs, vaccines, or surgery. This silly logic neglects the countless people, particularly children in India, who have died from diseases such as measles or even chronic diarrhea for lack of proper treatment. When you place your trust in a proponent of Ayurveda, you are also placing your trust in someone who likely claims to be able to levitate, read minds, foretell the future, reduce crime and end war through meditation, or heal with chanting, cow dung, and spit. Ayurveda is such a nonsensical practice—such a throwback to ancient, magical healing—that it is bewildering that most of its Western followers are well educated. India has largely abandoned Ayurveda, opting for vaccines and water treatment facilities. Only the poorest in India are stuck with Ayurvedic therapies. Only folks in the richest nations in the world embrace Ayurveda. The bottom line is that none of these New Age health gurus lives any longer than the general population. Perhaps irony, and not Ayurvedic forces, is the mysterious energy that governs us all.

29

৺ ৺ ৺ ৺ ৺ ৺ ৺

Something Smells Funny:
Aromatherapy as a Cure

Odors can indeed be powerful. The smell of urine and garbage can incite a riot. Citronella repels insects. But can aroma cure disease? Absolutely not. This is where aromatherapy goes astray. In terms of healing, a scent can only help you relax, if that.

One problem with aromatherapy, as many serious aromatherapists will attest, is the fact that it is so broad. Shaving creams, lipsticks, eye masks, candles, oils to rub, oils to burn, oils to drink . . . this is all considered aromatherapy. Marijuana smoking can technically be called aromatherapy, too. Another problem is certification, which can be attained, for a cost, in as little as a week. This lowering of the bar allows too many quacks to call themselves aromatherapists. Some of these certified aromatherapists have zero concept of the science of odors, let alone a concept of the scientific process in general. They merely collect and market aromas said to do this or that based on testimonials, sort of like trading recipes.

Thus, the first thing you'll notice in aromatherapy is that many of the oils do not do the things they're supposed to do. Claims are purely anecdotal and founder upon closer scrutiny. For example, three essential oils prescribed for alertness—peppermint, jasmine, and ylang-ylang—have been shown to be no more effective at increasing alertness than water. It seems that the anticipation of a smell and the knowledge of the effect it is *supposed* to have will influence a person's reaction to it. The smell of lavender can excite or place the body in a state of relaxation, depending on what you think it can do and what you want it to do.

The bizarre premise of aromatherapy is that essential oils are the spirits or souls of the plants that produce them. They can restore balance to your body, a common theme in alternative medicine based on yin and yang (and even on medieval Europe's four humors). Aromatherapy's "doctrine of signatures" maintains that a plant's look and smell dictate its medicinal effect. Mean-looking plants are powerful; violets appear gentle, therefore violet oil produces a calming effect. This is reminiscent of the homeopathic idea that snake venom, which is paralyzing, will clear up that stiff shoulder. There's a certain caveman logic to it, too. "Sun make fire. Me make fire. Me sun." Also, aromas are said to elicit emotions, which cannot really be qualified or substantiated. Sandalwood will boost your confidence; peppermint gets rid of negative energy; and patchouli makes one lust for peace. Ironically, patchouli grows wild in the Middle East.

Little science is done to see how chemicals in the aroma affect the nose and lungs and trigger the creation of molecules within the body. In the professional aromatherapy journals, you will see such language as "So and so prescribes such and such oil for x, y, and z. She also says it is useful for a, b, and c, as well." There is never any mention of scientific tests to see if these claims are true. The tests would be rather easy to perform. Just line up two groups, one getting a whiff and the other getting something else. Why isn't this done? Because aromatherapists, with their weeklong training, simply do not think this way. They have absolutely no concept of how the scientific process works. Pick up any aromatherapy book or professional journal. The omission is striking. The books *Essential Aromatherapy* and *Essential Oils* are good examples.

In these books and others like them, you will see something that is almost scientific . . . almost. First, there is a clinical description of the plant with its Latin name. Then there's the history. Oil from this plant was used in the Middle Ages to do this or that. Native Americans used it to treat x and y. You start thinking to

Aromatherapy cures are based on the alignment of planets. Want your jasmine to work? That depends on what Saturn is doing. *Courtesy of NASA/Hubble*

yourself, hey, this oil sounds like it could be useful. Next, if this is one of the "better" aromatherapy books, you'll read the ingredients. The oil contains certain percentages of alcohol, ester, things you never heard of, and maybe a vitamin or two. Next, you'll read about ways you can prepare it—perhaps burning it or rubbing a few drops into whatever body part ails you. Finally, you'll read about more benefits of the oil, usually something modern and timely like asthma, computer eyes, or carpal tunnel syndrome. Nowhere do you get any notion that this oil has been tested to be effective in treating the ailments mentioned above. All you have, really, is the testimony of superstitious medieval Europeans who thought that the Great Plague was caused by someone looking at you the wrong way. The occasional "scientists say" or "researchers have shown" only implies that the other 99 percent of the listings have absolutely no scientific validation.

The more you delve into aromatherapy, the more you will learn that this is a fruity world of nonscientific folks captivated by potpourri. Nearly all of the books are written by women without medical training. Men go along with aromatherapy because it's something the wife wants them to try that's not too much of a hassle, as opposed to giving up beer or taking up exercising. Aromatherapy also has sex appeal, with oils and massages and candles and soft music.

Aromatherapy need not be a joke. Ingesting certain oils either through the skin or under the tongue can certainly have an effect on the body. Yet this is less aromatherapy than it is herbal therapy. Vapor rubs open up nasal passages. Yet this is less aromatherapy than it is vapor-liquid medicine. Marijuana is an odor that will certainly induce a state of mind. Vapors from certain chemicals can kill you. The nose is a legitimate passageway for chemicals to enter the body. Yet aromatherapy as it is practiced today does not concern itself with delivering medicine through the respiratory system. Aromatherapy is only concerned with pretty smells, essentially perfume.

The greatest mystery is why Americans choose to believe one aromatherapy cure but not another. Aromatherapists could never sell eye of newt to rid a patient of rickets, as they did centuries before. That's just too silly for serious, health-conscious Americans. But we will burn peppermint to remove negative energy and restore harmony. The same folklore is behind both cures. Aromatherapy remains a pseudoscience chiefly because the aromatherapists themselves—lost in a realm of anecdotes, energy flows, harmony balancing, and superstition—have yet to establish a system of tested cures with safe doses.

30

☤ ☤ ☤ ☤ ☤ ☤

Suffocating Trends: Oxygen—
How Much Is Too Much?

From the smelly air of aromatherapy we jump to pure oxygen. Getting extra oxygen, O_2, is a peculiar new health trend, available from sources such as bottled water with dissolved oxygen atoms and oxygen bars that offer their clientele expensive ten-minute whiffs of pure or near-pure oxygen gas (as opposed to the 20 percent we get from air). The actor and noted hemp enthusiast Woody Harrelson opened an oxygen bar in Hollywood. I know what you're thinking: "Woody Harrelson! I loved him as a bald-headed mass murderer in *Natural Born Killers*. Surely he knows a thing or two about my respiratory system." Well, believe it or not, Woody might be a little off this time.

The notion that we need extra oxygen is ludicrous. Patients dying in hospitals are sometimes gradually given more and more oxygen as a last-ditch effort to save them because their lungs are incapacitated. Too much oxygen can be dangerous, though, and doctors can only keep patients in this state for a short time. Oxygen can be toxic to the blood. Adults with emphysema, chronic asthma, or chronic bronchitis cannot handle pure oxygen because it causes them to retain too much carbon dioxide. Premature babies who are given extra oxygen because their lungs aren't sufficiently mature to transfer oxygen into the blood can go blind if the concentration gets too high, a malady called retinopathy of prematurity, which Stevie Wonder experienced. Also, oxygen may be what ultimately kills you, rusting your body from the inside in a process called oxidation. The human body has adapted quite well to this lower atmosphere of ours that is roughly 20 percent oxygen, 75 percent nitrogen, and 5 percent trace gases. Blood cells, on exiting the lungs, are nearly saturated with about 97 percent oxygen bound molecularly to hemoglobin. Getting more oxygen serves no purpose.

Regardless, oxygenated bottled water cannot deliver any oxygen to you through your throat and stomach. It's biologically, physically impossible. Oxygenated water is one big scam, as explained below. Oxygen bars, which charge about a dollar a minute to breathe, are just stupid. Breathe, Inc., is but one of a few companies that sell oxygen machines to people wanting to strike it rich opening oxygen bars. Breathe, Inc., where "it's o.k. to inhale." They liked that line so much they got it trademarked. Selling oxygen, as you can imagine, must be tough. In fact, the Breathe, Inc., brochures attempt to coach potential oxygen bar owners through the tough questions they might hear from their customers or financial supporters. My favorite: "Is it dangerous to breathe Pure Oxygen?" The response: "Most doctors and recently the director of the American Lung Association said that there is probably no risk when used for short periods (less than sixty minutes) through a nasal cannula. Oxygen is and [sic] irritant to the lungs with long exposure (more than several hours)." This begs the follow-up question of why breathe it at all if it could ultimately be harmful.

At an oxygen bar, customers order small masks to cover the nose and mouth. The masks are connected by a hose to an oxygen tank. The thirsty customer pays by the minute, and starts a-huffin'. As always, we have the testimonials. Breathing pure oxygen will clear up your sinuses, make you think more clearly, make you more alert, help you catch your breath, cure headaches, and more. None of this has been proven true, even the part about catching your breath, despite what you see on *Monday Night Football*. Athletes would do just as well to sit on the bench and breathe deeply and slowly. If you want to think more clearly and cure a headache, maybe stop going to trendy, expensive bars with loud, inane pop music and fools breathing through masks.

The oxygenated water business is outright deceptive, and the FTC is starting to crack down. First there was Vitamin O, going for about $10 an ounce. What got one of O's marketers, Rose Creek Health Products, in trouble was its full-page ad in *USA Today* a few years back. Vitamin O is "stabilized oxygen molecules in a solution of distilled water and sodium chloride," according to the manufacturers. The ad contained those testimonials so familiar in the alternative medicine/snake oil world: more energy, better focus, never catch colds or flus. The ad also used a tactic common among oxygen sellers, suggesting (falsely) that the earth had more oxygen a few millennia ago and that air pollution was taking the oxygen away. What pushed the ad over the edge, though, was the statement that this oxygenated water technology was used for astronauts to ensure they have enough oxygen for their health in space. The FTC filed a complaint in U.S. district court against Rose Creek and a second marketer, The Staff of Life, which made similar unsubstantiated health claims. The court forced the companies to stop making false claims.

Yet here come more oxygen sellers, selling oxygenated water as a sports drink. The water will get oxygen back into your system and get you playing at peak performance again, the claim goes. Smart Water will help you think more clearly. And the Millennium Oxygen Cooler packs in 600 percent more dissolved oxygen than ordinary water, according to its ad. Of course, most of this dissolved oxygen will bubble out when exposed to room temperature and pressure, but why blind these companies with silly physics? The bottom line is that humans cannot absorb oxygen by swallowing it. Not even fish can do this; they filter water through their gills. One deep breath will bring in several orders of magnitude more oxygen than drinking oxygenated water. Robert Park of the American Physical Society estimates that you would have to drink a liter of the oxygenated water every 25 seconds to get a 1 percent oxygen boost, and this assumes you don't pee. If a friend is still convinced oxygenated water contains usable oxygen, suggest that he submerge himself in it. I doubt he will be able to breathe under water.

How can companies even market such products? The Dietary Supplement and Health Education Act of 1994 states that "natural" products do not need to be tested for safety or effectiveness. The FDA can pull such a product from the market only if it starts causing health problems after the product is on the shelves. Vitamin O is salt water. This cannot harm your body, only your pride and your pocketbook. The burden is on the FTC to monitor for false advertising and file complaints accordingly.

One intriguing topic in the oxygen world is ozone, O_3. There has been lively debate in the science community over whether filtering ozone gas through the blood of a living person can cure cancer or AIDS. For cancer, this is based on the false notion that cancer cells proliferate in a low-oxygen environment. Thus, cancer would choke if flooded with ozone. The theory is wrong, thus the treatment is useless—despite fringe cancer-patient advocates who insist the governments of the world are conspiring to keep this cheap ozone therapy out of the mainstream because it would undermine the financial livelihood of hospitals and drug companies. Ozone, they claim, is a panacea. Although ozone can kill bacteria in water supplies, the gas isn't much help in the body.

Ozone can kill HIV, the AIDS virus, in vitro or in a test tube. Could it work in the body, in vivo? Sadly, no. Doctors are still tweaking the process with hopes of finding an AIDS cure with cheap and easy ozone. German doctors are performing the bulk of this research. The idea is not so far-fetched, but at the same time it is not too promising.

31

ЁЁЁЁЁЁ

The Ultimate Hands-Off Approach: Touch Therapy, *Qigong,* and *Falun Gong*

I was given a troubling assignment from the *Washington Post* health editor in the summer of 2001. I was asked to report on the possible health benefits of *falun gong,* a set of five exercises practiced by followers of *falun dafa,* a spiritual movement banned in China. Many, including the Chinese government, consider the movement to be a cult. I needed to report on the exercises themselves, not the politics of them. I needed to be descriptive and objective, yet at the same time I could not encourage readers to participate in falun gong. Suppose this thing really is a cult? My fear was quickly diverted toward a much larger worry: university-based researchers funded by the National Institutes of Health were getting tax dollars to investigate magic; and the *qigong* health expert on the White House Commission on Complementary and Alternative Medicine Policy said she has healed people with her thoughts by phone as far away as Germany from her practice in San Francisco.

My research into falun gong first led me into the world of qigong. Qigong (pronounced chee-gung) is a modern adaptation of a three-thousand-year-old Chinese healing art, which includes tai chi and acupuncture. Millions of Chinese practice qigong exercises outdoors, often at sunrise. The trend started in the 1950s, and the newly established Communist party didn't seem to mind. The exercises can be slow and graceful or resemble rigorous calisthenics. Tai chi is a martial-art form of qigong; acupuncture is qigong with needles. The idea here is that through the exercises or stimulation with needles you can direct the flow of *qi*—also called *chi,* vital energy—to parts of your body that need it the most. More on this later. Falun gong is a qigong upstart developed by Hongzhi Li in China in 1992. I could find no medical expert, the kind needed for a newspaper interview, who knew enough about falun gong to comment on it. My experts could only comment on qigong, instead.

Qigong certainly has its place in the medical world. Since 1990, the National Institutes of Health have funded several small studies on the effects of qigong exercises for sufferers of neurological disorders, arthritis, and other ailments. One study funded by the National Institute on Aging found that folks over seventy years old gained strength and cut their risk of falling by nearly one-half after practicing tai chi. Larger studies are in the works. Acupuncture seems relatively sound as well, delivering at least some relief for nausea and pain.

One researcher I spoke to explained that science isn't so interested right now in how qigong might work but rather in *if* it works. Studies are simple enough. Group A does qigong exercises; group B doesn't; you measure the difference between the two groups after six months, and so forth. Perhaps folks who are older, frailer, or otherwise incapacitated can benefit greatly from a concentrated, nonvigorous exercise like tai chi instead of L.A.-style step-aerobic-kick-boxing. Meditative qigong can induce what is called the relaxation response, which is the opposite of the fight-or-flight response. Qigong-induced relaxation can decrease metabolism, lower the heart rate, and enhance resistance to disease. Saying the rosary can do the same thing. The fight-or-flight response, which you'll encounter if someone pulls a knife on you, quickens the heart rate and prepares the body for immediate danger. This is useful, but this is not a mode you want to be in all the time, because this kind of stress fosters disease. If tai chi, acupuncture, and meditation—collectively known as internal qigong—prove useful, it is likely due to the gentle movement or the relaxation response. There's no magic here.

External qigong is the stuff of magic. This includes touch healing, in which the healer touches a person's body or keeps his hands a few inches above it. Distance healing is the extreme—healing a person in Europe from your office in the United States. Any legitimate qigong master, either Chinese or American, will tell you that external qigong grew out of Chinese folklore. External qigong offers no chance of healing. The touch therapy idea came from Buddhist monks about two thousand years ago who, after meditating long hours, experienced a great warmth in their hands. The warmth came from an increased blood flow to the hands, away from the core, for the body enters into a relaxation state during meditation and does not need to protect vital organs by retaining warm blood in the center of the body. Ordinary folks, aware of the monks' warm hands, asked to be touched and healed, for the warmth seemed so radiant. Indeed, a warm hand feels nice on an upset stomach.

If you have ever seen a Chinese martial-arts movie, you know that powers tend to be exaggerated. Fighters leap atop houses and perform multiple flips in the air, all the while tossing knives and whistling the theme to *The Bridge on the River Kwai*. Touch therapy is just as macho. Soon, the stronger monk healers didn't need to touch to transfer the heat. The very strong could stand six feet away and heal. Ordinary folks, desperate for the strongest touch therapists, opted for the ones who could heal at greater and greater distances. And a pseudoscience was born.

The NIH National Center for Complementary and Alternative Medicine (NCCAM) is funding studies of touch therapy to see if it works. While testing all forms of alternative medicine is a noble goal, perhaps certain boundaries should be drawn. These could be boundaries perhaps set by the laws of physics, or internal qigong instead of external qigong. In the NCCAM-funded study, researchers in Michigan are monitoring the possible benefits of touch therapy on wound healing, which attempts to remove negative energy and instill positive energy—whatever that means. The touch therapist won't touch the wound, which is fortunate, because the chances of infection are far greater than the chances of magic healing. One must wonder, however, if any of these researchers read the *Journal of the American Medical Association* (*JAMA*). A few years ago, nine-year-old Emily Rosa devised an experiment for her fourth-grade science class to test whether twenty-one touch therapists had any ability to sense the "energy" of her own hand. The touch therapists merely had to determine if their magical hands were hovering over Rosa's hand. In 280 tries, they scored 44 percent—worse than a 50–50 guess. Rosa's experiment was so sound that it was written up in the *JAMA*. She became the youngest scientist to be published in a peer-reviewed medical journal.

Times are tough, and prestigious universities desperate for cash have become frantic for its funding. What's a school to do, other than adopt the name of a rich family? When news spread that NCCAM's budget grew from about $2 million to $100 million from 1990 to 2002, everyone began to salivate. Maybe, researchers began to ask themselves, we can do one of these easy studies where group A gets happy thoughts and group B gets negative thoughts. It'll pay the bills. We can only hope that the bulk of NCCAM funding is going into something legitimate and promising, such as testing the effects of the herb black cohosh on menopausal hot flashes. After interviewing folks for the falun gong news article, I had to wonder about this. One NCCAM-funded qigong researcher explained how humans have lost their harmony, or their sense of qigong, over the years. Cats still have good qigong, he said. That's why they can jump from high heights, land on their feet, and still survive. (Cats survive because their internal organs are cradled in the rib cage; humans have everything dangling. Also, cats in the wild, free from humans with their shelters and Western medicines, only live for three or four years.) Another researcher, Effie Chow, who sits on a White House commission on health, told me she has healed patients from a distance without meeting them. She has also helped paraplegics sit up and move; and she said she could help Christopher Reeve, the *Superman* actor paralyzed from the neck down, if only she were allowed access to him. Why not heal him from a distance?

It gets worse. Falun gong combines the most foolish aspects of touch therapy, mind healing, distance healing, and qigong. The gist is that a practitioner cultivates an intelligent, golden-colored entity called the falun, which resides in one's gut in a fourth dimension and spins continuously, absorbing energy from remote regions of parallel universes to heal the body. The falun gong founder, Hongzhi Li, now reportedly in exile in Queens, New York, maintains that David Copperfield has some serious falun that allows him to walk through walls and perform magic. This is all in Li's book, a wild read. Master Li and other falun gong masters give you the falun. You are then good to go as long as you keep the bugger spinning. So as not to discredit tai chi and other forms of qigong, the falun gong energy comes from outside . . . far, far, far away. Qigong cultivates energy within the body, merely redistributing it through breathing, movement, and focus.

Falun gong comprises five exercises, four standing poses and one sitting, meditative pose. Can these exercises provide any health benefits? The falun gong practitioners, naturally, believe so. They tout everything from cancer remission to an end of chronic diarrhea. Complementing the exercises, though, is the concept of *xinxing,* a code of morality one must practice or else the exercises are said to have no benefit. Abiding by xinxing also separates falun gong from qigong. The purported benefits of falun gong play out like late-night television testimonials. One retired white-collar worker from Beijing told me that falun gong had cured his skin allergies and chronic diarrhea. (He also joked about failing the xinxing morality test.) A Chinese woman spoke of how falun gong helped her regenerate bone, which had been removed in surgery. These practices can be mentally dangerous, though, when they instill false beliefs that lead to bad decisions.

Movements like falun gong enter into the realm of quackery when they consistently make health claims that cannot be verified scientifically. This includes healing by touch, raising the paralyzed, curing cancer at far higher success rates than conventional medicines, sending vibes across the sea to heal at a distance, or living healthy to the age of two hundred or more—normal aspects of falun gong. Bad decision making enters the picture when the more passionate of practitioners refuse medication in favor of falun gong. Hongzhi Li, the founder, clearly states that practitioners will never get sick if they properly cultivate the falun. Taking medication implies that one does not believe in the falun, thus illness becomes a test. You can criticize China for cracking down on falun gong, but don't get too caught up in political correctness. Falun gong is weird stuff; practitioners have set themselves on fire thinking they will be spiritually healed.

Will NCCAM fund research into falun gong? Why not? Folks practice it on the National Institutes of Health campus, and NCCAM seems to be funding everything else. Where do you draw the line in the wacky world of traditional medicines when the researchers are either plain wacky themselves or are hopping on the gravy train of research money for therapies they quite possibly know won't work? Right now, the approach seems to be: test all therapy, regardless of whether it breaks the laws of physics. When researchers find a 10 percent or 20 percent positive effect from a clearly groundless therapy, they know it is from the placebo effect. So what is next? Do we advocate the placebo effect as therapy? Do we mislead the public into thinking that a touch therapist is making their wounds heal faster? What possible outcome will arise from the "test all therapy" approach to NCCAM funding? May the Force be with us.

32

🜍 🜍 🜍 🜍 🜍 🜍 🜍

Getting to the Root of the Problem: Herbs As Alternative Medicine

At long last, we come to an "alternative" that has potential. Herbs hold great promise in the treatment of disease and everyday aches and pains. At least a quarter of the pharmaceuticals sold in the United States are derived directly from plants. Aspirin is a synthetic version of a compound found in willow tree bark. The only problem with herbals is that we don't fully understand which are good for what conditions and, more pertinently, what the proper doses are. The real downside is that the Food and Drug Administration does not regulate herbals. What is written on the label and what is stuck inside the pill don't necessarily jibe. Herbs vary wildly in terms of overall content, contamination, and plant part.

Verro Tyler, professor emeritus of pharmacognosy and natural drug products at Purdue University, wrote the bible for discerning the what, when, and how of the herbal world. It is *The Honest Herbal*; the fourth edition was published in 1999, and perhaps in a few years an updated edition can include results from recent NIH-funded herbal medicine studies. Tyler explains how the herb milk thistle, long thought to help the liver, has indeed been shown in clinical studies to protect against hepatitis and cirrhosis. Milk thistle scores a 95 percent success rate for curing acute poisoning from eating death-cup mushrooms. It is *the* medicine, not an alternative, for death-cup poisoning. In Europe, herbals such as milk thistle are regulated to ensure proper dose and purity.

Tyler also explains how the herbal field is undermined by untrained herbalists, aromatherapists, astrologers, and New Age healers who blindly recommend herbal remedies with no clue to how dangerous they can be. Tyler's clinical reference book has trouble competing against the flashy, upbeat herbal medicine guides that contain nothing but fluff. These modern herbalists make big claims that have not been proven. Does St. John's wort really treat mild depression? Maybe not. Does saw palmetto really prevent prostate cancer? Probably not. Does ginkgo really boost your memory? Most likely not. The National Institutes of Health's National Center for Complementary and Alternative Medicine has been looking into these claims. NCCAM's approach is that if Americans are going to consume these herbs blindly, we should at least have reliable evidence that they are safe, effective, and available at proper doses.

There have been some winners: black cohosh, as herbalists have said, seems to control hot flashes and other symptoms of menopause *as effectively as or even more effectively than* conventional treatments. And there have been losers: blue cohosh, to the surprise of traditional healers, seems ineffective at treating menstrual cramps and may be a tad poisonous. Black cohosh has been said to cure a lot of things, including snakebites, according to certain Native American tribes. These are the types of practices that herbalists, like aromatherapists, swap like recipes. Except for relieving hot flashes, none of the other claims for black cohosh are true.

It shouldn't be hard to see how natural herbs can be dangerous. Poison ivy is natural, but it is not something you would use in a skin cream. Mushrooms are natural, but half the species can kill you. Very potent yet common herbal remedies include mistletoe, comfrey, and foxglove. These can all be deadly at even moderate doses, and you may be unaware of the true dose within each pill. So the claim that herbals are safe because they are natural is clearly false. The most toxic compounds known, such as strychnine or amatoxin, are derived from plants. Many herbals also cause allergic reactions. The most common are chamomile, echinacea, and feverfew, all of which can be troublesome to people allergic to ragweed, chrysanthemums, daisies, and other flowers in the aster family. Because the FDA does not regulate these remedies, you may not find such information on the herb's label.

Some herbals are just shoddily made. They contain either junk ingredients, nonpotent sections of the plant, tainted ingredients, or sometimes no herbal medicine at all. Your chances of knowing this are slim. A few privately funded organizations, such as http://ConsumerLab.com, regularly test herbal products and publish lists of those that make the grade (thus avoiding lawsuits by *not* publishing products that *don't* make the grade). Good quality control is rewarded with a certification that we can only assume is legitimate, although there is no quality control of the privately run quality controllers. These third-party organizations do frequently find that a product claiming to have 5 percent of a given herb might have zero percent, 0.001 percent, or 99 percent. So it can hurt you or do nothing. This is often poor quality control and not deliberate fraud. Some herbals, particularly from Asia, may also contain harmful metals such as mercury. This won't be on the label, even if you read Chinese.

How can anyone get away with this? Well, it all goes back to the Dietary Supplement and Health Act of 1994. The act treats herbal supplements more as a food than a drug, merely requiring manufacturers to ensure that a supplement is generically safe before it is marketed. The FDA steps in if it learns that an herb is not safe or if manufacturers make medicinelike claims. Until this law changes, herbal remedies can contain just about anything, and the manufacturers can claim just about anything. Sure, the act prevents herbal supplement labels from using the words *cure, treat, mitigate,* or *prevent* when referring to a disease. This is simply the background music for the herbal two-step, in which manufacturers dance around these words to describe their products. Ginkgo biloba "promotes mental alertness" instead of "treats Alzheimer's disease." Ginkgo commercials sport a good-looking man in his sixties saying he's worried about losing his memory and wants to take precautions; Alzheimer's is implied. Ginkgo has not been shown to improve mental alertness, let alone reverse Alzheimer's symptoms.

A drug maker could never get away with such a claim. Is our sixty-year-old ginkgo-popper taking aspirin to prevent heart attacks? We hope not, because ginkgo has anti-clotting properties, just like aspirin and other blood-thinning drugs, and combining the two kinds of substances increases the risk of internal bleeding and strokes—although again, the average consumer will not know this.

Now some companies are putting herbals directly into food and drinks. Not only don't Americans understand that herbals can be dangerous but also we treat them like candy. You can buy sugary teas, sodas, waters, and snack bars with dozens of different herbal medicines. This is the stuff that makes Verro Tyler and serious herbalists cringe. It cuts to the heart of what is wrong with the American approach to medicine.

Americans, upon hearing that an herb or vitamin may be good, will go out and buy it in great quantities, in its highest concentration, and incorporate it into the sickly American diet. For example, recent studies have shown that green tea may prevent breast cancer. The American researchers said that although their study doesn't prove that green tea prevents breast cancer, it isn't harmful and there's a good chance it can be beneficial to health. Therefore, they recommended that Americans drink more green tea. This is the worst recommendation the researchers could have made. Americans aren't going to drink green tea the way the Japanese drink green tea. The Japanese drink it straight. Americans don't like green tea. It's too bitter. They will only drink "green tea drink"— a concoction of water and green tea extract, plus sugar and salt— the same junk that makes Americans unhealthy in the first place. The same goes for ginseng drinks. They are just sugar-sodium water with, if you are blessed, a touch of ginseng, although minus the active ingredients, which are too expensive to include. (Only roots at least four years old contain useful amounts of the active ingredient.) Likewise, what good is the stress-relieving herb kava-kava when it is bound to a chocolate bar? The edible-herb trend merely promotes unhealthy eating in a way that makes you feel good about it.

No book on bad medicine and bad herbs would be complete without mentioning the product Bloussant, which is, according to WellQuest International, Bloussant's manufacturers, a "less invasive alternative to cosmetic surgery" for the ever-popular quest to have "increased cleavage, firmness and fullness." This product is aimed at teenage girls. The ads, often quarter-page size, are in teen magazines such as *Seventeen,* and there are commercials for it on cable television as well. Bloussant is a mix of don quai, black cohosh, fennel seed, and saw palmetto. How much of each is anyone's guess, because, by law, the manufacturers do not need to divulge that information. How does it work? Bloussant will "wake up your body's growth process" and "actually stimulates the inner-cellular substance in the breast," the magazine ad says. "Your confidence level will soar! Until now, our only options were to just live with small breasts, use artificial padding, or endure very expensive & dangerous surgery." It's good to know WellQuest is looking out for the health and confidence of young women. This is the state of the $31 billion dietary supplement industry.

The bottom line is that herbs, like everything else, are made of chemicals. Some chemicals are very safe for humans; some chemicals are very dangerous. It doesn't matter whether man or nature synthesizes the chemical. It's still a chemical. There is no logic in the idea that nature's chemicals are safer than a pharmaceutical company's chemicals. Thus, ingesting an untested herb is no different from ingesting an untested pharmaceutical. Furthermore, no medicine is inert. Medicine is effective only when it changes something in your body. Medicine that works is, by definition, a chemical that is potentially harmful to your body over time.

33

ዩ ዩ ዩ ዩ ዩ ዩ

A Shot in the Arm:
The True Dangers of Vaccines

The fear of vaccination is tinged with sadness. Each vaccine does indeed come with a risk. There is a relatively high risk, perhaps one in several thousand, depending on the vaccine, that a vaccinated child will become ill with a high fever or flulike symptoms. This could be from an allergic reaction. There is an incredibly low risk, less than one in several million, depending on the vaccine, that a vaccinated child will grow seriously ill and die, or become brain damaged or paralyzed. This small risk, understandably, runs through every parent's mind when she subjects her child to a vaccine. Low risk does not mean no risk.

A vaccine is a dead or severely weakened virus that in its active form would invade the body and cause disease. Your body can easily conquer this compromised form of the virus. Your immune system remembers what the virus looks like, and, in the future, if you come in contact with a meaner version of the virus, your body will have the internal biological weapons to fight it. This is called immunity.

Viruses are the simplest of life forms, ten to a hundred times smaller than bacteria. (Bacteria such as anthrax can also cause disease, but they are usually treated with antibiotics; the term vaccine *usually* applies to viral infections.) Viruses comprise only a protein shell and a simplified version of DNA. They need other living cells—ours!—to reproduce and spread. Colds and flus are caused by viruses. Same goes for AIDS, diphtheria, polio, smallpox, chicken pox, whooping cough (pertussis), lockjaw (tetanus), measles, mumps, rubella, and more. Much of the human race, from its dawn well into the twentieth century, has died from a viral infection.

Through vaccination, viruses can be rendered extinct. Smallpox and polio have almost disappeared from the world. The trick is to immunize everyone so that the virus has nowhere to live. This is called herd immunity. All we need is a good generation or two with most people immunized, and the virus will be gone. Viruses, unlike bacteria, serve no benefit to humans and can do only harm. Virus extinction is a good thing.

When a parent denies a child a vaccine to a now-rare disease such as polio, that child is not in immediate harm. Polio is extremely uncommon in the United States and Canada, so the chance of that unprotected child contracting polio is rare. If you take a trip to New York City or some urban melting pot, though, your chances of encountering the polio virus shoot way up. The Caribbean experienced a polio outbreak in the 1980s; Haiti and the Dominican Republic experienced a smaller outbreak in 2002. Anyone living in or visiting that area was a potential carrier. The world is getting smaller, and the risk of contracting a virus from another part of the world is all the greater. The real crime of denying polio vaccination, however, is the selfish undermining of the world's effort (and this *is* a worldwide effort) to eliminate the scourge of polio. The folly of denying a child a vaccine for much more common diseases—for example, measles, mumps, and rubella—is far more apparent. Thousands of unvaccinated children die each year or develop brain damage after coming in contact with these viruses. In Afghanistan alone, the World Health Organization estimates that 35,000 people die of measles each year. This is the world without vaccines.

An antivaccine underworld—very well armed with "facts," I might add—has spun a series of myths about vaccines that has convinced many well-educated, nature-loving types to forgo vaccination. Some propagators of the myths are, tragically, parents whose children have been harmed by vaccines. Others are conspiracy theorists with a bunker mentality, convinced that the CIA is controlling the population and that fluoride in drinking water is a Communist plot. Their most vocal argument—that the measles-mumps-rubella, or MMR, vaccine causes autism—experienced a few glory days in the press but has since been proven to be utterly false. Other arguments are largely misconceptions about viruses or simply falsehoods.

The MMR-autism scare flared up in 1998 when the researcher Andrew Wakefield published his study results in *The Lancet,* a prestigious British medical journal. Wakefield had no intention of scaring anyone. He merely reported that autism and gastrointestinal problems were associated with (that is, appeared at the same time as) the MMR shot, which is typically given during a child's first year of life. Although from a scientific point of view the association made little sense (no biological mechanism, no animal studies, no convincing statistics with human cases), the claim was a sensation, and the well-educated, wealthy, *au naturel* crowd ran with it. The very low if not improbable risk of developing autism from an MMR vaccine, based on a casual observation in one contested study, was apparently enough to place children at a very real risk of contracting measles, mumps, and rubella. These diseases can kill babies or leave them with learning disabilities.

Wakefield's study was interesting and worth following up. That came—lots of it, in fact. The U.S. Institute of Medicine (IOM), an independent, nongovernmental group of the brightest American medical researchers, concluded after several years of study that there was no connection between autism and the MMR vaccine. Yes, the MMR vaccine was introduced as late as 1963. (And it lowered yearly cases of measles from about 500,000 to 500 by the year 2000.) And yes, the rates of autism do seem to be rising. We just happen to be diagnosing autism at that stage in a child's development when its body is ready to handle the MMR vaccine. Autism is a condition that likely begins before birth, not after. The IOM concluded that autism is likely caused by a combination of genetics and environmental factors, which may include prebirth diseases and compromised immune systems, but not the MMR vaccine itself. The MMR issue came to the forefront once more in early 2002, when Great Britain's prime minister, Tony Blair, told the public that his newborn child had indeed received the vaccine—and assured his nation of the vaccine's importance and safety. Blair also denied the rumor that had been circulating months before that he didn't trust the vaccine for his own child. Nevertheless, organizations in Britain such as JABS (Justice, Awareness and Basic Support), a support group for vaccine-damaged children, remain vociferous in their stance that the MMR vaccine is dangerous and unnecessary.

Sudden Infant Death Syndrome, or SIDS, also seems to occur at about the same age as many routine infant immunizations. The issue has been studied extensively, and there is no evidence of a cause-and-effect relationship. The possible connection, from a grieving parent's point of view, must seem overwhelmingly convincing. Here, a mother sees her baby getting shot after shot and test after test during that first year. Then the innocent, apparently healthy baby inexplicably dies while sleeping. There is a great need to place the blame somewhere.

Now for the conspiracy arguments. A prime one claims that polio was dying out naturally. The polio vaccine gave polio to hundreds of thousands of children, and this is the sole reason why polio exists today. Polio rates were indeed declining through the end of the nineteenth century up until the late 1950s, when Jonas Salk introduced the first of two main polio vaccines. The rates were declining because of cleaner water and better personal hygiene, a triumph of the public health movement. Polio is transmitted through fecal matter. You'd be surprised and disgusted at how easily transmittable polio is. One common transmission ground was public swimming pools. Hygiene could help only so much, though. The vaccine has nearly wiped out polio worldwide. But polio rates fluctuate from year to year. The anti–polio vaccine stance is that the rates jumped 5 to 500 percent in the years after the vaccine was introduced. There indeed may have been jumps in local polio rates here and there in the 1950s. This is natural. If one person contracted polio in a small town one year, and, in the next year, an epidemic hits and infects 5 people, that's a 500 percent increase. Perhaps many more would have been infected in this town had the townsfolk not been vaccinated.

The addendum to this argument is that the oral polio vaccine—which was developed by Albert Sabin and contained a weakened live virus as opposed to Salk's injection of a dead virus—infected over a hundred school children in the 1960s. No one knows how to answer this claim. Polio vaccines were a tough sell. Imagine convincing skeptical parents back in the 1950s and 1960s that the polio vaccine their kids were drinking contained a live polio virus. Polio vaccines were mandated by law. Accusations of jeopardizing the health of the nation flew wildly, even in the halls of Congress. So, when polio struck—and it did, because we hadn't yet established herd immunity—the oral vaccine was the first to be blamed, followed by the injected vaccine. The success of the polio vaccine, fifty years later, is obvious, but the early days were filled with justifiable fears.

The oral vaccine does lead to a polio infection in one in 2.4 million doses, researchers have since learned. The unfortunate victims were infants with abnormal immune systems and a few adults who had never been immunized for anything. A safer, enhanced, inactivated polio vaccine (eIPV) has since replaced the oral vaccine in the United States. To say that doctors are murdering innocent folks in developing countries by administering the seemingly more dangerous live, oral vaccine is an insult to the dedicated souls saving millions of lives each year. The live vaccine actually works better in regions such as sub-Saharan Africa. Health experts are unable to vaccinate everyone there; the weakened live vaccine, like the polio virus itself (and unlike a dead vaccine), can spread from one person to another, thus providing protection to the nonimmunized.

Other antivaccine folks echo such faulty thinking as: vaccines are more dangerous than the disease itself; there's no strong link between vaccination and disease decline; and, just how effective are vaccines anyway? All such claims have been proven false, with deadly consequences. One argument suggests that the odds of having a reaction to the vaccine for whooping cough, or pertussis, is about one in two thousand; yet the risk of dying from whooping cough is about one in a million. This sounds like a convincing argument for dropping the vaccine, but both sets of numbers aren't what they appear to be. That one-in-two-thousand risk corresponds to the likelihood of a mild, nonfatal reaction. The risk of *dying* from the vaccine is far lower than one in a million. The risk of catching and dying from whooping cough after declining the vaccine seems so low because everyone around you has the vaccine and doesn't spread the disease. Go to a country where whooping cough is prevalent—or come in contact with a person from that country or anything he touches—and you just may get whooping cough and get seriously ill from it.

The pertussis vaccine has been a common target. European studies found that pertussis deaths were low across Europe regardless of the rate of pertussis immunization from country to country. Here we see the herd immunity in effect. A lot of vaccinated folks were protecting those without immunity—so many were immunized that very few were actually carrying the disease. Yet Great Britain decided to let its guard down, save some money (taxpayers' money, antivaccine people like to argue), and cut back on pertussis immunization. The vaccination rates fell in 1974, and by 1978 there was an epidemic of more than 100,000 cases of pertussis and 36 deaths. Sweden and Japan went through the same thing.

You want a cause-and-effect association? With the breakup of the Soviet Union, diphtheria immunization fell by the wayside. The number of cases rose from 839 in 1989 to nearly 50,000 in 1994, including 1,700 deaths, according to the U.S. Centers for Disease Control and Prevention. Diphtheria spread into Europe and the United States as a result. Time after time a cutback in immunization ushers in a revival of disease. On the flip side, we have success stories. Since the introduction of the Hib *(Haemophilus influenzae type b)* vaccine in 1990, Hib rates have fallen 99 percent in the United States.

Those who argue against vaccination have, unfortunately, lost the sense of how miserable life was before vaccines—how entire families were wiped out, how a husband and wife had ten children knowing that seven would likely die before reaching adulthood. If I haven't convinced you of the importance of vaccines, visit the Immunization Action Coalition web site at http://www.immunize.org. You'll read testimonials from parents whose children suffered or died from diseases that could have been prevented by a simple vaccine.

PART VI

ꝗ ꝗ ꝗ ꝗ ꝗ ꝗ ꝗ ꝗ ꝗ

Risking It All

*The desire to take medicine is perhaps the greatest feature
which distinguishes man from animals.*
— Sir William Osler (1849–1919)

The ongoing joke (or frustration) among researchers in the risk sciences is how people sweat over little risks while engaging in high-risk activities. The examples are endless. We ski, snowboard, and participate in extreme sports, yet we pass legislation to ban raw-milk European cheeses because of the six people who die each year after eating them. We demand that the Environmental Protection Agency lower the pesticide levels in foods because of a false perception of a rise in childhood cancers, yet we do nothing about the fact that thousands of children are killed each year by guns, either by accident or by intent. We demand that the air be cleaner to minimize the risk of lung disease, yet at least 25 percent of the population in most countries smoke cigarettes. We worry about the lifetime risk of death from pesticides (at the 1-in-1-million level) yet not about the lifetime risk of death from driving (at the 1-in-100 level). Over three-fourths of all heart disease, strokes, and incidences of diabetes, along with many cancers, can be avoided or significantly delayed for decades with modifications in diet and exercise. These are the big killers, the real risks.

34

꜒꜒ ꜒꜒ ꜒꜒ ꜒꜒ ꜒꜒ ꜒꜒ ꜒꜒

Toxic Avenger:
The Science of Toxicity

Everything is toxic. It's true. Even water is toxic. Too much of it will kill you. It's called drowning. Toxicity depends on the dose. Too much of anything—salt, oranges, dioxin—will cause harm. Toxicology attempts to determine what levels are safe.

Industry produces some deadly solvents and byproducts, such as dioxin, benzene, and vinyl chloride. We commonly call these chemicals "toxic" and indeed they are, but only at certain levels. A solution of one molecule of benzene mixed in a billion molecules of water is not toxic. In fact, it is completely harmless. One in a million is pushing it, though. Admittedly, benzene has the ability to be toxic, or poisonous, at lower concentrations than, say, ethyl alcohol (liquor) or sodium chloride (salt). Benzene is over a million times more toxic than ethyl alcohol. Yet how many more people die from alcohol poisoning each year compared to benzene poisoning? A whole lot more, you may have guessed. So which is the more dangerous chemical? "The dose makes the poison," said Paracelsus, a Swiss physician of the sixteenth century and the father of toxicology.

The trick is determining how much is too much, the toxicologist's burden. A toxicologist determines a safe level of a chemical by conducting a series of studies on rodents; noting lethal, sublethal, and reversible doses; noting whether the substance is causing cancer, mutation, nerve damage, or other irritations; tracing the chemical's final destination and noting whether it is excreted or exhaled or stored in fat or bone cells; taking into account hyper- and hyposensitivity; extrapolating to human body weight, exposure rate, exposure pathway, and life span; and assessing the necessity of the chemical, be it for medication (high risk allowed) or indus- try (high risk frowned upon).

Complicated? Now we see why industry sometimes acts as if dioxin is safe enough to add to your breakfast cereal, while Green- peace and ardent environmental groups claim there is no safe level at all. It's all a matter of how you interpret the animal and cell data, and these studies are tricky. For example, humans can easily tolerate ingesting copper at a few parts per million. This level is deadly to algae. (Yet algae need copper at the parts-per-billion level to reproduce!) Animal studies on copper's toxicity might not tell us much about what humans can handle.

Dioxin is a known animal carcinogen. A mere microgram of the substance (a billionth of the weight of a raisin) per kilogram of body weight is enough to kill half the guinea pigs exposed to it. By comparison, it would take a milligram (1,000 micrograms) of nico- tine or 100 milligrams of DDT to kill that many rats. Nasty stuff, dioxin. Is it harmful to humans? Definitely, at high levels. Is it harmful to humans at the levels found in the environment—con- centrated in meat and dairy products, including Ben & Jerry's ice cream? That's a complicated question.

The Environmental Protection Agency does not yet have the data to place dioxin in the "known human carcinogen" category. Instead, dioxin is called a "probable human carcinogen," causing, perhaps, about an additional five hundred cancers a year. Consid- ering that the sun causes more cancer than dioxin, it is difficult to support the claim that dioxin is "the most toxic substance known to man," as is often repeated. Botulinus toxin, the cause of botu- lism, is at least a hundred times more toxic than dioxin, molecule for molecule. Hemlock and blowfish toxin aren't so nice, either. And, as we know, alcohol kills far more than all the chemicals defined as "highly toxic" combined, save nicotine.

Yet we cannot let dioxin off the hook, for dose makes the poi- son. The problem with dioxin is twofold. First, it hangs around. Most of the dioxin comes from burning plastic and other organic material in trash, forest fires, and automobile tailpipes. The chemical doesn't break down easily in the air or soil but rather settles "as is" on and in blades of grass. Cows eat the grass and concentrate the dioxin in their fat cells. So beef and milk, which are loaded with fat, contain concentrated levels of dioxin. The fattier the product, such as Ben & Jerry's ice cream, the more dioxin. We humans love fat. Thus, we eat the dioxin from the trash we just burned a few years ago by eating cows and drinking milk, and we store this in our own fat cells. The dose, albeit small, starts to add up, because we store concentrated dioxin.

The other thing about dioxin is that, while the chemical may or may not be causing human cancers, it may be responsible for other health and environmental problems. Fish, amphibians, and reptiles—from frogs in the Great Lakes region and alligators in the Everglades—seem to develop inadequate genitalia or even to change sexes from dioxin pollution in waterways. The data are not conclusive, but it is certainly reasonable to be alarmed. (Fears that dioxin is causing lower sperm counts in men and raising the rates of testicular cancer and female breast cancer, however, seem unfounded.) Toxicologists process all of this information. They know that human cancer from dioxin isn't obvious yet, but just as algae die from a slight increase in copper concentrations, humans may suffer from higher levels of dioxin. Toxicologists assess the pathway and elimination route: it's concentrated in food, and there is zero elimination. They then factor in the necessity of dioxins. Plastics, the source of most dioxins, are important. Bleached white paper, the source of most of the other dioxins, isn't so important. The good toxicologist hands a report to the EPA, and the EPA makes a decision.

The EPA—still unable to classify dioxin as a known human carcinogen, like alcohol, arsenic, and asbestos—has forced the industry to cut dioxin emission by more than 90 percent from 1980 levels as of 2002. Whereas municipal waste incinerators emitted about 18 pounds of dioxin yearly through the 1980s, the level will drop to only half an ounce per year. Medical waste incinerators will drop from five pounds of emission to a quarter of an ounce. This will dramatically minimize our yearly dose of dioxin. Greenpeace, whose hard work and, some say, relentless bullying tactics influenced the EPA's decision, will likely continue fighting the paper mills as well as society as a whole for wasting so much paper—the bleaching of which adds dioxins to the environment.

The EPA is less concerned with pesticide residue on food. Here we find the dose is low, however toxic the pesticide might be. Most pesticides wash off, and the pesticide residue that resists peeling and cooking in the kitchen does not accumulate in the body the way that dioxin does. The importance of pesticides is considered high compared to that of dioxin, for it is arguably more difficult to grow food cheaply and plentifully without pesticides. Also, some cooked foods contain natural cancer-causing agents more dangerous than pesticide residue. That crispy black coating or searing on a catfish or steak is a known human carcinogen, far worse than pesticides or the small amount of dioxin you might be ingesting.

Some so-called toxins, such as copper, are necessary for life. Selenium is a trace mineral found in a variety of soils and in the plants grown in those soils, such as wheat. Too little selenium in the diet, less than 20 micrograms a day, can lead to thyroid problems and Keshan disease, which is characterized by an enlarged heart and poor heart function. This is a problem in parts of China and Russia, where selenium levels in soil are naturally low. Yet selenium, unlike other nutrients, has a very narrow window of opportunity, and too much (perhaps as little as a thousand micrograms each day) can lead to massive organ failure. Some selenium supplement pills come close to this toxic level. Wheat grown in Nebraska up through the Dakotas and central Canada is naturally rich in selenium. As mentioned in chapter 21, vitamins A, C, and E are all beneficial to health but become toxic when taken in high doses. One aspirin can cure a headache; the whole bottle will kill you.

The take-home message is that every substance on earth is toxic at a given level. There are absolute safe levels, at which a small amount of a potent poison will have zero effect on the body. And there are also acceptable levels, depending on the necessity of the chemical. How many people would die, for example, if chlorine *weren't* added to drinking water to kill bacteria? The answer is tens and even hundreds of thousands each year.

No one claims that toxicology is a precise science. Consider that Albert Einstein proved Newtonian physics to be imprecise. Newton's physics was good enough, though, to determine the orbits of the planets and lots of other nifty stuff. This is where we are with toxicology, getting a loose idea of what's bad and what's all right. Unfortunately, industry hides behind this imprecise science, and the government is not always clever or diligent enough to protect us. The classic case concerns asbestos.

Health experts spent decades trying to ban the commercial use of the fibrous asbestos mineral while industry hid facts, censored health studies, and flat-out lied in court—reminiscent of the tobacco industry. As related humorously in Sheldon Rampton and John Stauber's 2001 book *Trust Us, We're Experts!*, companies elbow-deep in the mining and manufacturing of asbestos played a game of denial and blame shifting—slowly admitting first that asbestos could be dangerous to workers, but not their particular kind of asbestos or not at their exposure rate, then that the exposure rate could cause breathing problems but not cancer, and so on. This culminated in cries that banning the deadly asbestos would cripple industry and the United States itself. Looking back, it is now clear that some companies knew how deadly asbestos fibers were to their workers but did nothing to protect them. Studies were manipulated to show Congress that asbestos wasn't causing silicosis and lung cancer in young workers in their twenties, thirties, and forties, when all medical researchers knew that lung cancer has a twenty-year dormancy period and only the older workers (who were not examined) would be adversely affected by long-term exposure.

The same sad tale is told for lead and other industrial toxins. The removal of lead from gasoline was a monumental victory over two of the biggest industries in the world, oil producers and automobile manufacturers. Again came the denial, followed by arguments about lead being bad but not *that* bad; then the complaints about crippling industry and America's prosperity. Currently, the chlorine industry—from an outside perspective—appears to be acting like the other industries before it. The old Monsanto motto, "Without chemicals, life itself would be impossible," seems sinister now, a public relations campaign that backfired and turned as comical as the "This is your brain on drugs" commercials. Chlorine is not deadly, the industry says. Studies are inconclusive, they say. The risk is overstated, they say. America's prosperity is at stake, they say. All of this might be true. Given American industry's track record of denial and cover-ups, however, one can understand the worry of Greenpeace and other environmental and health groups. The EPA, perhaps, took a preemptive strike in reducing dioxin emission.

Although the general public seems relatively safe from industrial toxins, the same cannot be said for the worker. Only within the past few decades have workers been protected against industrial toxins such as vinyl chloride and even dioxin, which causes immediate and often permanent skin rashes at high exposure rates. For decades the coal industry knew how deadly coal mining was but refused to supply protective equipment to its workers. Most uranium miners, too, were never protected and never compensated. Lead, mercury, tin, arsenic, and nickel were all common elements that workers were exposed to regularly at high doses with no protection, from the founding of the United States well into the 1960s. U.S. industry has, for the most part, cleaned up its act—although nonunionized, migrant workers from Mexico and Central America are regularly exposed to dangerous pesticide levels during spraying on farms from Texas and California up through the Great Plains. Sadly, in U.S.-owned companies in developing countries, anything goes. Workers are not given protective clothing, let alone ventilators. This is why it is cheaper to do business outside of the States.

Yes, there's chlorine in drinking water, dioxin in ice cream, benzene at the gas pump. They are toxic only at or above certain levels. When it comes to toxins, I personally fear less for the general public and more for the unprotected worker.

35

Peer-Reviewed for Your Pleasure:
How Health Studies Work

Trying to make sense out of the latest health findings can be frustrating. Studies released only months apart find contradictory results. Eggs are bad for you. Eggs are good for you. Brown eggs laid on Fridays are better for you than white eggs laid on Mondays if you're a postmenopausal Asian woman age fifty to sixty-five. And so on. Who's conducting these studies? Why can't they get their act together?

There are four main reasons why, year after year, health studies seem to contradict each other. The first is bias. Sometimes scientists have an unconscious agenda; or, interested parties can manipulate studies so that the results make them look good. The second reason is study strength. Big studies, which usually yield more statistically sound results, cost big money and are not often performed. Also, it's not always possible to conduct a long, comprehensive study on, say, pesticide use and the risk of developing cancer in twenty years, when health and economics are at stake now. This leads to quick and inexpensive studies and studies of varying strengths, and targets may vary in results. The third reason is the way the study is reported or interpreted. A newspaper may report the entire finding, but you might only catch the headline, which really doesn't capture the true medical findings. The fourth reason is one that people, even doctors, seem to forget: boy, is the human body complicated.

Health studies are not meant to be conclusive. Nor is peer-reviewed publication a validation of a study's conclusion. The editor and jurors of a peer-review health journal simply make sure that the science performed in the study is relatively sound. Doctors will gladly admit this. The purpose of these studies is to gain insight, to pick up a few clues. The language of subsequent peer-review–published reports merely states that substance or behavior X was shown to produce a Y effect in Z percent of the people, or rats, in the study. Few studies show that X causes Y. This is hard to relate in newspapers, though. For example, the result of a rat study that finds that caffeine raised the level of certain chemicals in the bloodstream associated with higher cholesterol levels, which in turn is associated with circulatory disease, translates in the headlines as: Coffee may cause heart attacks. The intricacies of the study are usually left out of the headline and first few paragraphs of the story. This isn't the fault of lazy or dumb science reporters. Reporters merely give you a general idea of the study up front. If you are indeed interested in the story, you will likely read further about how the study was with rats, not humans; on caffeine, not coffee; on indicators of circulatory disease, not heart attacks. You may read that follow-up studies are planned for coffee, provided that the doctors secure money for coffee mugs tiny enough for the rats.

A few months later you may read that coffee is good for the heart. What gives? This might be the result from a study similar to the rat study, only now we are talking about humans drinking coffee. These subjects may drink, say, three cups of coffee a day and have a little bit of their blood drawn at the end of each day for two weeks. This blood is compared to blood from similarly sized or active people who don't drink coffee for two weeks. At the end of two weeks, the first group had slightly higher levels of a chemical associated with lowering cholesterol levels. Tediously, we can say that in two small groups of people, those who drank coffee for two weeks had a small but statistically sound rise in a certain chemical that, according to other published reports, may lower cholesterol when found at even higher levels in the body. Or, in headline fashion, we can say that coffee is good for the heart.

The researchers in study two are likely to have read the journal report of study one and said to themselves: Hey, we can conduct a better study. Or maybe they had heard about study one at a scientific conference and already had study two in the works by the time study one was finished and published. This is how the scientific process works. Researchers learn from the strengths and weakness of other studies and strive to conduct and publish better studies. Their careers—in terms of tenure and future funding possibilities—depend on this. Neither study is conclusive. We're talking about indicators of indicators of heart attacks. These studies can come across as being conclusive, but to the researchers the studies are just teasers—a way of testing the water to see if it is worth the time, effort, and money to perform a larger study on coffee consumption and circulatory disease.

By this point, the researchers have established a mechanism. Clearly, coffee drinking is not related to bunions, cold sores, or baldness. There's no mechanism. But the researchers have found that something in coffee—maybe caffeine, maybe not—is interacting with chemicals in the body to produce a notable change in the bloodstream. More small studies follow in humans (after all, coffee isn't a poison and we don't need to be pumping rats with a tall café latte from Starbucks). All the new studies seem to negate that initial rat study and find that coffee is raising levels of a chemical that helps lower cholesterol. With each new study result, we are being fed in the newspapers that coffee is good for the heart. Now it's time for a big study. Researchers want to enroll 5,000 adults in a five-year coffee study to see whether the coffee drinkers have fewer heart attacks compared to the noncoffee drinkers in the study. Will the National Institutes of Health pay for this? Not this time. So the researchers get money from the coffee industry instead. This may lead to biased reporting, but maybe trouble won't percolate. We'll see.

Five years later we learn that the noncoffee drinkers had fewer heart attacks. Here come the headlines: Coffee drinkers at greater risk of heart attacks. This is what the study found, but is it true? We still don't know if *coffee* caused the risk. The body is so complicated that coffee drinking may have no effect on it, either good or bad. Researchers must ask themselves what led to these results. Maybe the noncoffee drinkers exercised more or drank green tea instead, which may prevent heart attacks. Maybe the coffee drinkers smoked or, less obviously, were so stressed out and run down at their jobs that they needed to drink coffee to stay awake. Competing, scientifically minded researchers will feel obligated to duplicate this study with controls for all the factors in the first study (exercise, stress, green tea, diet in general) that may have influenced the results. Time passes. More studies. More opportunities for bias, yet ever closer to some kind of statistical truth.

With each subsequent study, we, the lay audience, get the gist of the result, usually in some kind of bold-type, thumbs-up or thumbs-down format. Newspapers and magazines, although ridiculed, are actually a decent source of health news if one is willing to read an entire article to see what a given health study really said. And of course the reader is welcome to read the actual journal article, in which the procedure and results are reported in painstaking detail. Television news, unlike the print media, often falls short, with the entire health study captured in only a few sentences. Try timing these stories on television sometime. You'll see that some last as little as ten seconds. They are artful in their brevity but not necessarily scientific.

So, does coffee cause heart attacks? Sorry, I have no idea. Coffee studies are ongoing. This is the nature of science, too: to perform studies that require more follow-up, and thus more funding, in order to stay employed. Homeopathy studies are notorious for this. Here we have a medicine that is only water, so all researchers are testing are the effects of water on disease. They compare homeopathy to a placebo. Naturally, the results are inconclusive, because they are measuring the effect of one placebo versus another. Sometimes homeopathy looks good; sometimes the other placebo looks better. In every homeopathy study there is always the conclusion that more research is needed.

The coffee industry has never been accused of manipulating coffee health studies. They do actually fund the Coffee Institute at Vanderbilt University. I'm sure the researchers there do good, honest work. Other industries have not been so honest. You know about the tobacco industry. These folks suppressed the negative studies that they funded and only published the ones that showed smoking wasn't harmful. Lung cancer takes around twenty years to develop, so it is easy to craft studies that show that smokers in their twenties were just as healthy as, if not healthier than, non-smokers in their twenties. What ultimately doomed the tobacco industry was the fact that lung cancer was so rare before the smoking craze. By the mid-1950s Americans were living long enough and smoking plentifully enough to contract lung cancer. So the tobacco industry went into denial mode, then manipulation mode, and then they just lied.

So did the asbestos industry, as discussed in chapter 34. It is a loose federation of mining companies, manufacturers, and even automobile and oil companies that use asbestos, a fibrous mineral with a thousand and one applications. When inhaled, though, the tiny fibers penetrate deeply into the lungs and cause asbestosis, a chronic inflammation and hardening of lung tissue. The industry knew of the hazards and purposely crafted studies that showed how certain types of people were not getting asbestosis. Thus, unbiased medical-school studies were saying asbestos was bad; and biased, industry-funded studies were saying asbestos was fine. Headlines in newspapers went back and forth, and the lay public never knew what to think. Some say the chlorine industry is playing the same game with studies on dioxin—which some say is safe and others say is more deadly than the sweat of Satan—but only time will tell.

Sometimes health effects are just too complicated to determine, even with a battery of studies involving tens of thousands of people. They sound simple enough: give group A some beta-carotene and group B no beta-carotene and see what happens after five years. Or look back: examine cancer or heart patients and determine which took a vitamin and which didn't over the last five years. This is called epidemiology. But is five years long enough? Some say the body needs a lifetime of vitamin supplementation to truly ward off cancer and heart disease. Are 10,000 people enough? If the effects are minuscule or diluted by other factors (stress, exercise, diet, access to health care, mental attitude, family support, and many more), then the study would need more people to draw more statistically sound conclusions. Even the location of the study—Europe versus the United States, say—will affect the outcome, even though the same types of people are receiving the same types of doses. The researchers' own honest bias is a factor, too, when they want the study to turn out a certain way. This is why the Linus Pauling Institute always seems to publish studies showing the benefits of vitamin C while other scientists cannot find such benefits. The Institute represents a continuation of Linus Pauling's latter-day legacy in the field of vitamin C research.

Epidemiology is an imprecise science, but it is the best we've got. With a combination of studies all looking for more or less the same outcome, scientists get a feel for what is going on. The public often doesn't have the patience for this. We want to know now, yes or no, whether vitamin E prevents heart attacks. With our current technology and tools of detection and analysis, we simply cannot know the answer for sure. People are more complicated than genetically identical rats eating the same food, living in the same environment, and running on their little wheels for the same length of time each day. Thus, human health studies take years of repetition and analysis before reasonable conclusions can be made. We are still in the middle of this antioxidant mess, so we go back in forth in the headlines on whether vitamins A, C, E, and selenium are any good in unnaturally high doses.

Should any of this matter too much? Common sense will tell you that moderation is the best policy when it comes to diet and health. Americans are often too quick to jump on the latest fad—the all-egg diet, the no-egg diet, megadoses of antioxidants, green tea, ginseng, fish oil, wheat germ. Let the scientists battle it out over what is beneficial to human health. You won't die from waiting. If, remarkably, a certain food or drink adds years to your life, the answer would pop out immediately. Stick to a lifestyle that has been recommended for ages—no smoking, low-fat foods, plenty of vegetables, and some exercise—and you cannot go wrong. If, in a few more years, researchers are convinced that the occasional Guinness stout is good for your health, then I'll drink to that.

36

꿈 꿈 꿈 꿈 꿈 꿈

Candy Adds Years to Your Life: And Other Important Health Study Findings

Experiments that seem the most useless to the general public are often crucial for the scientific process. Not all science involves a glamorous search for a cancer cure. Mostly it's grunt work, figuring out which chemicals interact with what cells in which organs of what animals. This provides the foundation for the science superstars who, for example, might determine whether supplements of vitamin E are advantageous to health. Scientists need to know, first, that vitamin E pills aren't dissolved and rendered worthless during digestion. Hence, the hardcore science literature hosts reports with titles such as "Alpha-tocopherol resilience in mammalian gastric acid solutions." Here, the scientist is determining whether the chemical component of vitamin E (alpha-tocopherol) breaks apart and converts to other chemicals when soaked in stomach acid. This is not a glamorous study, and to the outsider it may seem worthless. Why is this mad scientist dissolving chemicals in acid, we ask? Yet, all higher-level vitamin E epidemiological investigations depend on this study.

The real mad scientists are those who chose to test the outright silly. Will a kiss under mistletoe chase away a Christmastime cold? Can Beanie Babies fight depression? These are actual studies from scientists at reputable universities. They are essentially selling their services to industry to get some money to perform useless science so that the makers of Christmastime mistletoe can advertise that their product wards off colds. Scientifically proven! As seen on TV! Industry will fund a stream of studies until one, by chance, shows a positive result—in this case, that a group of people with mistletoe hanging in their houses experienced fewer colds than a group of people who didn't have mistletoe. Is there some chemical in the mistletoe that is protecting us from the cold virus? Probably not. Is the health study a good one? Definitely not.

You yourself can evaluate health studies in the safety of your own home. Hill's Criteria of Causality, developed by Sir Austin Bradford Hill and first presented in 1965, comprises a list of checks to determine the strength of a health study. One recent study from Harvard, on candy eating and longevity, thoroughly flunks Hill's criteria. The results were published just before Christmas in 1998 in the *British Medical Journal*. The journals do not censor science results, however dumb; they just ensure that the science isn't bogus. The line between bogus and dumb, though, can be razor thin. Let's examine this study.

The premise is that eating candy may increase your chance of living longer. The study looked at the lives of 7,841 men, free of cardiovascular disease and cancer, all alumni of Harvard who entered the prestigious university between 1916 and 1950. They were part of a decade-long research project on health and lifestyle. The candy groups were separated into nonconsumers (who answered "almost never" on the questionnaire) and consumers (who ate as little as a few pieces of candy a month or several a day). The questionnaire about candy habits was conducted in 1988; the deaths (514) were counted in 1993. Those who ate candy lived eleven months longer, on average, than those who did not.

Now for Hill's criteria. What's the strength of the association? Or, how strong was the effect? The paper reports a 0.92-year increase in longevity from eating candy, which is not much compared to the gain from eating lean meats and vegetables. The relative risk was 0.73, meaning the men in the candy group were only 27 percent more likely to live longer, which is weak by statistical standards in a small-sized study. Score, D+. What's the dose response? Does the more candy you eat translate to more months you can live? And, does sucking, eating, sugar content, or candy type matter? No. Actually, the group with the highest consumption of candy had about the same death rate as the nonconsumers. Score, F. What's the consistency of the response? Has this been reported before? No. Score, C+ instead of F, for the slim chance that this could be a pioneering study. What's the relationship between exposure and effect? Did the candy eating start very early in life or later? Not established. Score, F.

How specific is the chemical effect? That is, what does it do? The answer is that candy makes you live longer. This is somewhat ambiguous, not specific. Score, C−. What's the biological plausibility? Can you explain the results at the molecular level in terms of sugar and cellular metabolism? No. Score, F. Does cause-and-effect conflict with established knowledge of disease? Yes. Score, C+ instead of F, again, for the slim chance that this could be a pioneering study. Is there experimental evidence on lab animals? No. Score, F. (Remember, we need scientific grunt work to support human health studies.) Are there analogies? Do similar chemicals lead to similar effects? No. Score, F.

At this point, the careful reader might be inclined to say: Lighten up. This is just a silly and harmless study, right? Yes and no. The authors of the candy report certainly have participated in stronger work, and they approach this study, it seems, with tongue in cheek. They are by no means being paid by the candy industry to publish such results. All in all, the study is kind of fun. Fellow researchers got a good chuckle, no doubt. I use it merely as an example. But what's the message here? That candy is good for you? The real story is that sugar consumption in the United States is unfathomably high, about twenty teaspoons daily, according to the U.S. Department of Agriculture. Sugar is in our food and in our drinks. One can of soda pop contains about ten teaspoons of sugar, and sugar consumption is intricately tied to obesity and diabetes.

The authors cannot be faulted for how the news media reports study results, but here's what happened: On December 18, 1998, just a week before Christmas, one Philadelphia-area paper ran a fun, relatively long article about the study from the Scripps Howard News Service, which means that small-town papers around the United States most likely carried the same story. We enter into the holiday food orgy with the notion that Harvard scientists say sweets make you live longer. Bad advice. On December 31, when everyone was preparing to get drunk, the *Philadelphia Inquirer* ran a three-sentence news brief about how diets high in sugar lead to a variety of ills. Such news kind of dampens the holiday spirit, so it was downplayed.

We don't know why the men in the Harvard study lived an average of 0.92 year longer if they ate candy. It's likely that candy had nothing to do with the result. The results can be explained by chance or even lifestyle. Maybe the men who lived longer were "young at heart," and this was reflected in their desire to suck on candy, a "young" thing to do. Maybe the men had family members in their lives who brought them candy as a sign of affection. In this case, it's family support, not candy, that leads to longevity.

Canada, America's sane neighbor to the north, is not immune to silly health studies. Health researchers at the University of Toronto found that Oscar winners live longer than nonwinners. This was in the *Annals of Internal Medicine* in 2001. The scientists went so far as to say that winning an Academy Award instills in the recipient a sense of accomplishment and peace of mind, which could explain the extra longevity. Dumb, dumb, dumb. You are bound to find some difference between winners and those who are nominated and lose. Maybe one group has more broken bones. Maybe one group has more sisters who die of cancer. There will always be something. Can you attribute all of these to winning an award? Of course not. There is no logical connection. With the baby boomers in the United States and Canada reaching retirement age, however, we are consumed with longevity issues. Hence, this silly study is conducted and published. In reality, Oscar winners live no longer than the general population when adjusted for wealth and access to health care. Pumping up the numbers, no doubt, was George Burns, who lived to 100 and didn't win an Oscar until he was 75. So any scholarly explanation supplied to explain the health benefits of Oscar winning gets washed out.

Such silly studies serve as instructive examples of how bad some of the "serious" studies on homeopathy and dietary supplements are, even when they originate in such prestigious universities as the almighty Harvard. But what I want to know now is, how long do candy-eating Oscar winners live?

37
⚕ ⚕ ⚕ ⚕ ⚕ ⚕ ⚕

We're #1: Rating America's Health

s America #1? Maybe in basketball. The U.S. health care system ranked 37th on a list of 191 systems compiled by the World Health Organization in 2000. (France topped the list; most of Africa finished in the bottom third.) This doesn't mean health care is bad in the United States. It just means that 36 other countries have it better, on average. This includes Japan, Canada, most of Western Europe, and part of the Middle East. The really nice thing about American medical care is its ability to treat diseases and injuries that no one else around the globe can. The United States has diagnostic technology and surgical skills unmatched in the world. U.S. doctors perform intricate transplants and radical new surgical procedures on the brain, eyes, and heart on a daily basis. The sick and wealthy from the five other continents often fly to the United States for such procedures. Likewise, Johns Hopkins Medical School Hospital in Baltimore is the best hospital in the United States and arguably in the world. The Harvard Medical School area in Boston hosts several world-class hospitals that are the envy of the international medical community. Philadelphia, too, is home to highly respectable hospitals, such as the Temple University system.

So why does the United States rank thirty-seventh in health care, nineteenth in life expectancy, and twentieth in infant mortality? The problem, it seems, is a lack of preventive medicine. Every other industrialized nation has close to 100 percent of its population enrolled in a health insurance program. The United States has

217

about 60 percent insured. Also, the United States educates its public less about health (exercise, diet, sex) and provides fewer basic
necessities for its citizens: food, shelter, vaccination, and family
planning. Add to this the high number of homicides (30,000 annually, three times more than second-ranked Finland) and the highest
teen pregnancy rate (twice as high as second-ranked Great Britain),
and it is easy to see how the average person in America has it
worse than the average person in other industrialized nations.

Complicating issues is the fact that the United States has three
distinct populations: the very wealthy, a somewhat insured middle
class, and the poor. The poor are not just from the much-discussed
inner city. Large tracts of the United States—from Native American reservations and the Appalachian Mountains to rural regions
throughout America—have a health care infrastructure no better
than that in many developing nations in Africa and Central America. So, when a Native American woman develops breast cancer,
she usually dies. There are no regular breast exams. There are no
clinics for biopsies. Cancer is detected late, and the chance for survival (which could have been 90 percent in a better setting) is minimal. People in other industrialized nations would never find themselves in this situation. For middle-class America, chances are
much better but not superior to that of Europe and Japan. For the
wealthiest in America—who are often well educated and have
superior access to health care—the U.S. system is unparalleled.

Among other low points, the United States ranks fifteenth in
occupational deaths. An estimated six thousand workers are killed
by accidents each year and some fifty thousand die annually from
occupational diseases, according to the National Census of Fatal
Occupational Injuries. The United States doesn't do well by children, either. The country ranks first in child gun violence, first
among industrialized nations in the number of preschool children
not immunized, eleventh in the proportion of children living in
poverty (one in five), and seventeenth in low-birthweight babies,
according to the Children's Defense Fund. Children under age fifteen are twelve times more likely to die from gunfire, sixteen times
more likely to be murdered by a gun, eleven times more likely to
commit suicide with a gun, and nine times more likely to be killed

in a firearm accident than kids in twenty-five other industrialized nations combined. These sobering facts come from the Centers for Disease Control and Prevention. Also, as of 2001, of the 154 members of the United Nations, only the United States and Somalia have yet to ratify the U.N. Convention on the Rights of the Child.

The United States is #1 for a few things. In terms of health and the environment, the United States has the highest rate of beef and snack-food consumption and the highest rate of coronary bypasses; the highest rate of women who have had multiple abortions; the highest rate of HIV infection among industrialized countries; one of the worst ratios of doctors to patients and teachers to students; the highest rate of homelessness; the highest emission of air pollutants per capita; the heaviest garbage per capita; and the greatest disparity of wealth among industrialized countries.

The good news is that improvement need not be difficult.

PART VII

ꝰ ꝰ ꝰ ꝰ ꝰ ꝰ ꝰ ꝰ ꝰ ꝰ ꝰ ꝰ

Just Like in the Movies

I beheld the wretch—the miserable monster whom I had created.
—*Frankenstein,* by Mary Wollstonecraft Shelley (1797–1851)

Hollywood, as you can imagine, is rampant with bad medicine. Characters in the movies live and die in extraordinary ways. There is the obvious: Bullet wounds never become infected. You can knock a person out with one punch across the chin or one karate chop on the neck. Hollywood bodies clearly contain more than the normal five liters of blood in the human body, and Hollywood blood has the ability to squirt farther. Humans die on cue, after saying their final words. A chloroform-soaked handkerchief will instantly render someone unconscious. Bottles and chairs shatter easily over the head. Coma victims awake with perfect hair and makeup. Crowds in restaurants or on the streets never have handicaps (deaf, paralyzed, muscular dystrophy), a broken leg or arm, floating eyes, acne or skin rash, or cleft lip, and they are never pregnant, unless it's the pregnant woman who'll be kidnapped. No one uses a condom, and no one gets pregnant or contracts a sexually transmitted disease. Dogs can survive anything. Everyone, from Roman soldiers to medieval peasants, has perfect teeth.

There is also the not so obvious. In the real world, guns cause instant hearing damage; a whack on the head can cause a lifetime of neurological problems; and heart attacks often are not felt in the chest. Misconceptions perpetuated in Hollywood movies, in these cases, prove to be debilitating and sometimes deadly. Don't expect television news to set the story straight. Health and science reporting on television can be even more misleading than in the movies. It's a wonder broadcast journalists don't win Oscars.

38

☤ ☤ ☤ ☤ ☤ ☤ ☤

I'm Not a Reporter, But I Play One on TV: The Accuracy of Television Medical News

U nlike newspapers such as the *New York Times* and *The Washington Post,* which are expected to maintain a certain degree of journalistic integrity, television news has gone the way of pure entertainment. There's nothing intrinsically wrong with this. The problem is simply that most of us do not realize this is the case. We assume, logically, that what is on national network television during the news hour is accurate. A story that airs nationally on television, presented as news, is instilled with a certain level of legitimacy. Millions are watching. You would never expect mainstream newspapers or magazines to consistently feature sensationalistic science and health reports as if the topics were widely accepted. That stuff, you would think, belongs in magazines about UFOs or ghosts or paranormal experiences. Yet television news does. TV stations take the ghost stories, the mad-scientist stories, and the psychic stories and present them as news, because, let's face it, they are entertaining.

The CBS news program *60 Minutes* is cutting-edge journalism. The show has won many prestigious awards, and both print and broadcast journalists admire the show. It was a radical departure from television news and an immediate hit. That said, the show has featured some wacky health stories presented in a sensationalistic way. Many journalists would agree that *60 Minutes*'s report on shark cartilage was one of the journalistic low points for this otherwise noble news program. ABC copied the format with *20/20*—even the name is similar. Other feature-oriented news programs followed, each one leaning more toward entertainment than hard news. When cable television became mainstream—with its endless choice of the marvelous, mawkish, and mundane—network television took a belly punch. The challenge was to make news even more entertaining to attract viewers who could just as easily switch to cable without leaving the comfort of the sofa.

The following is a description of a health report that appeared on ABC News's *20/20 Downtown* on August 13, 2001. My intent here is not to pick apart ABC News, although admittedly I do just that. Rather, I believe the ABC report provides an instructive example of how bad medicine is presented as exciting news: with pseudosymmetry, or the pretense that most experts are in agreement with what is being presented; a dependence on unusual or sensational science results that others in the scientific community renounce as unsound even before the air time (and the reporter knows it); and a reliance on "experts" and an avoidance of impartial or critical voices. Yes, ABC has fine reporters. The medical editor Dr. Timothy Johnson is a valuable addition to the news team. And yes, ABC News does a fine job explaining new open-heart surgery techniques or the latest discoveries from NASA's Chandra X-ray Observatory. These are examples of hard news. Where television in general gets bad is when it comes to news *features* presented as hard news.

The topic of the *20/20 Downtown* report was distance healing and prayer. The reporter, Michael Guillen, holds a Ph.D. in physics from Cornell University. That's impressive; Cornell has one of the top physics departments in the United States. Guillen, the author of several books, is also the science editor at ABC News and has the stated goal of eradicating scientific illiteracy. So I was figuring I was in store for some top-notch, accurate, no-holds-barred reporting.

For several decades now, many scientists have been saying that religion can promote good health. Studies have shown that folks who attend church regularly, pray, or involve themselves in religious activities enjoy longer, healthier lives than those who don't. If true, this may be because these people get outside more often. Maybe they walk to church. Maybe they sweat and toil at church functions, such as bake sales or house-building. Maybe they simply remain connected to a society where people care for them and can look out for them if they need a ride to a hospital. There are likely many positive things deeply religious people do unconsciously for their health. Also, praying is known to trigger the relaxation response, lowering metabolism and heart rate and strengthening the immune system. Herbert Benson of Harvard Medical School is a leading researcher in the area.

Guillen presented a different kind of story. Guillen starts the piece by stating that 90 percent of us pray in times of hardship. He then jumps right into a series of images of people around the world praying for one complete stranger, a fellow undergoing heart treatment at Duke University Medical Center in North Carolina. This guy, who looks to be in his sixties, is part of a scientific test on the healing power of prayer. Doctors will determine if prayer—performed by Tibetan monks, American born-again Christians, and others—can affect the health outcome of this and other Duke heart patients. Straightforward enough; results are pending.

Guillen then informs us that 191 studies have looked into distance healing and the power of prayer, and two-thirds show "tantalizing" results. What studies, where? How positive? Guillen didn't say. But if the study he chose to highlight in the next scene was any indication, the word "tantalizing" doesn't mean "noteworthy." The core argument in Guillen's case for distance healing being real is a study performed at the Mid-American Heart Institute in Kansas City, Missouri. The study followed the fate of 1,000 patients admitted to this center's critical care unit.

Of 1,000 patients passing through the critical care unit, half are placed in a group that will receive prayers for a year from complete strangers, organizations not obviously affiliated with the health center. Half are placed in a group that doesn't get prayers. Both groups, of course, get the quality heart care that only the Mid-American Heart Institute can deliver. And no patient is aware that he is being prayed for. This eliminates the placebo effect. We don't know the selection process—that is, which critical-care visitors were excluded from the study—but this is an understandable omission by Guillen for the purpose of time. It is also unethical to enlist patients unknowingly in a "prayer group," because individual patients may have personal objections to being prayed for—either in general or by groups outside of the patient's own religion. This wasn't addressed, either.

The result? After a year, the group that was prayed for had 11 percent fewer heart attacks, strokes, and life-threatening complications throughout the year. Even if the study didn't have serious design flaws, an 11 percent difference in poorly defined health outcomes in a study with only a thousand patients is hardly something to get excited about. Chance alone could have produced those results. In fact, that's what follow-up analysis and published reports about this study stated. Guillen, one might assume, would have known of the overwhelming limitations and criticisms of this study. You don't get to be an ABC science editor, or a Ph.D. from Cornell, by missing the obvious. Nevertheless, Guillen repeats this 11 percent figure twice, stressing it to imply that 11 percent is a big difference. This study showing an 11 percent difference was, perhaps, the best that Guillen could find among the two-thirds of 191 studies that had "tantalizing" results.

I had many questions. If 90 percent of us pray, as Guillen stated up front, then surely those patients in the "no-prayer" group were really getting prayer, likely from loved ones at home if not themselves. There is a certain level of background prayer. If this is the case, then this would imply that simply praying a little isn't enough. The group without the benefit of strangers praying for them were, over all, more unhealthy, right? Perhaps you need lots of prayer—the kind that comes from Tibetan monks. Logically, it would follow that different levels of prayer may affect your health by varying degrees. Patients who had twice as many prayers could perhaps be twice as healthy.

So many questions, and Guillen cuts right to the chase. He looks the doctors involved in the study straight in the eye and asks: "Do you take this as evidence for the existence of God? It's the most obvious question." I was let down. I thought the most obvious questions were: Why do you think this study shows anything more than what can be explained by chance? Or, do you think health is affected by the number, type, frequency, or length of prayers, the distance from the patient of the people praying, or the religious affiliation of those involved? Or, how do the born-again Christians you enlisted feel about the prayers from the Tibetan monks, who will ultimately suffer a fiery eternity for praying to a false, non-Christian god?

Guillen didn't challenge the study at all with direct questions, as any good reporter would. The scientists who conducted the study, who for all we know might be very good scientists, are never placed in a position of defending what purports to be a groundbreaking revelation about energy fields unexplainable by the current laws of physics. We move to the next scene, where, after these doctors conclude that God must have willed certain heart patients to be healthier than others as a result of random prayer, we hear that "it's a miracle." Guillen is now talking to a psychiatrist from California. (You can always find professional support of the bizarre in California.) The miracle is not the Kansas City heart study, though, as we will find out in another minute or so. The miracle is an AIDS study similar to the heart study. Ten AIDS patients got the prayer treatment from a traditional healer miles away, and ten AIDS patients got no prayers. Four patients in the no-prayer group died, yet all the patients in the prayer group were still living. We know nothing about the study design or the background of the patients. This doesn't matter to Guillen; he's merely supporting the evidence shown in the heart study.

Next we visit the Ayurveda guru and best-selling author Deepak Chopra, who for the last fifteen years has forwarded the notion that the mind can heal the body and reverse the aging process. Chopra will be featured for two minutes and forty-five seconds, which is 27.5 percent of the total segment and far more than any other interviewee gets. Perhaps Chopra deserves this allotment, for after all, his book, *Grow Younger, Live Longer: Ten Steps to Reverse Aging,* is being released the same week this ABC program is airing.

What does all this mean, Guillen (a trained physicist, remember) asks Chopra, who has been introduced by this point as a health expert. Many in the health and science community question Chopra's "science" (he won the Ig Nobel quack award), but his background is not supplied. Chopra is quite clear in his answer: "What physicists are saying to us right now is that there is a realm of reality that goes beyond the physical where in fact we can influence each other from a distance."

Yes, there are forces in the universe we cannot explain, such as the theorized "quintessence" that seems to oppose gravity and accelerate the expansion of the universe. There are virtual particles in the subatomic realm that pop in and out of existence, according to the math of quantum theory. There also exists entanglement, what Einstein called "spooky" action at a distance, in which two particles once joined can affect each other when separated and placed miles (or light years) apart. But physicists are not suggesting that humans can heal each other at a distance. Chopra, the expert, says this; Guillen, the physicist-cum-reporter, doesn't challenge it; and the viewer is left to assume that this is scientific dogma.

Guillen next allows Chopra to demonstrate these mental powers. We are taken to the Institute of Noetic Studies in Northern California, "a place where scientists routinely test paranormal phenomena," Guillen says. Again, Guillen presents scientific oddity and unproven claims as everyday fact. The Institute of Noetic Studies, it seems, is Room 300 in a nondescript building. No university; this is fringe science, at best. Chopra hooks up Guillen to a machine that measures nerve activity. Guillen is asked to relax while Chopra enters a different room. Chopra views Guillen on a monitor and proceeds to use his mind power to make Guillen relax. After fifteen minutes of this, they analyze the data. A computer printout shows Guillen's nerve activity going up and down, as nerves do. We cannot see the scale on the graph, but at rest, nerve activity cycles as you breathe, swallow, scratch, or whatever. A second curve indicates the intervals when Chopra used his mind power to get Guillen to relax, showing that Guillen relaxed more when Chopra willed him to. Guillen claims that the curves overlapped. Even if they did, it would prove nothing, because we have no idea what this nerve machine does. The sad thing, however, was the fact that if television viewers could freeze the scene and zoom in on the chart, they would see that the curves didn't always overlap. Chopra merely got lucky a few times when Guillen was at a low point in the nerve cycle. Furthermore, the entire experiment is nothing more than hocus-pocus with fancy equipment. There is no reason for Chopra to view Guillen on a monitor if Chopra has the ability to calm him from a distance. The video monitor simply makes the experiment seem more scientific.

More talk from Chopra follows, and finally, about eight minutes into this ten-minute segment, we hear from a doubter. He is introduced as "skeptic Dr. Gary Posner." Guillen does not state Posner's degree or professional affiliation. The Posner section is a full forty-five seconds, longer than is normally allotted to a dissenting voice. Posner states what should be obvious to any medical professional watching this program: the heart and AIDS studies were greatly flawed, and distant healing studies in general are often shoddy. Viewers would not harbor seeds of doubt for long, though. We cut back to Chopra for a wrap-up.

Guillen literally asks Chopra to put a positive spin on Posner's negative interpretation of distance healing studies. This is the final fluff question in a fluff piece that failed to address serious issues, such as the fact that children of parents in certain fundamentalist Christian groups die because the parents refuse to give the child medical treatment and instead rely on prayer. A national health system based on distance healing would horribly undermine the great advances of twenty-first-century medicine.

If I could have asked one question, it would have been about quantum healing, a concept explained in Chopra's book *Ageless Body, Timeless Mind: The Quantum Alternative to Growing Old.* Maybe the healing we saw in the Kansas City heart study was the result of quantum praying? Stay with me now. A quantum is the packet of energy needed to change the energy level within an atom. Maybe prayer is useless against fighting disease until it reaches a certain capacity. Ten prayers are needed, and 9.9999 prayers won't do a thing. This is why the "no-prayer" group, which only got a few residual background prayers said for them at home, were less healthy. It took the extra prayers from strangers to reach the needed level of energy to trigger the healing process.

Joking aside, this is the state of television health news. I could have randomly picked any network broadcasting any health feature. (In fact, I did. I saw ABC's *20/20 Downtown* only because I happened to be rewinding a videotape.) Such a report may be entertaining, but it leaves people with a sense of false hope. False hopes lead to bad decision making. A week or so later, the *New York Times* ran an article about a paraplegic artist who has movement only in his eyes and eyebrows. Through an elaborate set of eye movements, the man commands his assistants to push his wheelchair, covered in paint, over a huge canvas. The story touched upon real issues, real technologies, and real hopes for other paraplegics and, when it comes right down to it, for all of us who feel we cannot make it through the day. Meanwhile, Guillen, who aims to eradicate scientific illiteracy, once again keeps us chained, not inspired, with a report on magic and fraud. This can only give the medical charlatans of the world the upper hand in pulling one over on the public.

39

Rambo VI: The Quest for Hearing: Guns and Their Aftereffects

New from Tri-Hard Pictures and Recycle Cinema comes a tale of courage, loyalty and determination . . .

Kompongcham, Cambodia—dawn

Close up on our HERO, dining on a breakfast of grubs and foul grass tea that is mercifully weak. The unrelenting heat and humidity choke the entire jungle, even at this early hour. The men are scared; the smell of danger hangs near.

SGT. CODDLE
We have to move out. Chopak's men are within a mile.

RAMBO
What?

SGT. CODDLE
(louder)
We have to move out.

RAMBO
What?

Coddle, a more patient man under normal circumstances, rips Rambo's note pad from his shirt pocket and writes out the message. Rambo nods.

RAMBO
Sorry, my ears are shot from all this gunfire.

Rambo reaches back to grab his rifle, a foolish move. His shoulder pops out of its socket, as it has so easily in recent years, the result of sustained shoulder dislocations from shooting massive guns with one arm. Rambo falls to the jungle floor, wincing in pain. Coddle knows the drill. He signals for two men to hold Rambo's body back and he stretches Rambo's arm back into the shoulder socket.

SGT. CODDLE
(mumbling)
I'm getting too old for this.

Did you miss *Rambo VI* in the theaters? Rambo is stone deaf, as you can imagine he would be after firing all that heavy equipment without ear protection. Exposing your ears to 110 decibels for a few minutes each day may cause permanent hearing loss. This is the level you'd experience at a music concert. In fact, Pete Townsend, leader of what was once billed as the loudest rock group, The Who, is almost completely deaf from all those years of performing concerts. Poor Rambo has endured far worse. The sound of gunfire echoing in a jungle canopy is easily thousands of times louder than a concert, enough to instantly cause severe and permanent hearing loss—especially because a gun must be held close to the ears to fire.

Just how loud is a gun? Riding in a convertible down the freeway or riding in a subway car is pretty loud. This is about 95–100 decibels. Keep going. Power saws and sandblasters are pretty darn loud too, 10 times louder than the subway at 110 decibels. Keep going. Car horns and jackhammers are at the threshold of discomfort, defined as 120 decibels, 100 times louder than the subway. Keep going. An air raid siren will shake you out of bed at a distance of 5 miles; at 130 decibels it's 1,000 times louder than the subway. Keep going. The sound of handgun and military assault rifle fire is at 140 decibels, defined as the threshold of pain, 10,000 times louder than the painfully loud sound of a subway. Keep going; Rambo is no wimp. Missiles and larger assault rifles, the kind that Rambo likes, fire at 150 decibels. Bazookas kaboom at 160 decibels. And the 105-mm Howitzer is a menacing 190 decibels when fired.

Table 2—Decibels. Say what? Gunfire is at least a billion times louder than an air conditioner and can cause instant, irreversible hearing damage.

Noise Level in Decibels	Example	Comments
0	Threshold of hearing, sound of one hand clapping	
10	Breathing, beginning of a Pink Floyd album	Just audible
20	A whisper, rustling leaves	
30	Quiet rural area, birds chirping	
40	Library, birds fighting	Quiet
50	Conversation at home (no kids)	
60	Conversation at home (with kids), restaurant, air conditioning, television, an office	Intrusive
70	Vacuum cleaner, noisy restaurant, telephone dial tone, middle of a Pink Floyd album	
80	Garbage disposal, typical factory, passing freight train, hair dryer, alarm clock, city traffic from inside your car	Still a million times quieter than a gunshot
90	Busy street corner, diesel truck, blender, train whistle at 150 meters, subway train approaching	Hearing damage occurs after 8 hours straight
100	Motorboat, lawnmower, leaf blower, motorcycle, tractor, riding in a subway, riding in a convertible	Serious hearing damage after 8 hours
110	Steel mill, power saw, car horn, rock concert, sand blaster, firecrackers, headphones (funneled right into your ear canal)	Still a whisper compared to a gun
120	Loudest human scream, thunderclap, chain saw, jackhammer, Spinal Tap concert	Human pain threshold; workers allowed only 15-minute per day exposure at 115 dB
130	Jet takeoff at 100 meters	
140	Aircraft carrier deck, air raid siren, handgun, military assault rifle	Instant hearing loss
150	Jet takeoff, missiles, large assault rifles	Eardrum ruptures, Rambo level
160	Bazookas	What?
170	10 bazookas	What?
180	Rocket launch pad	What?
190	105-mm Howitzer	What?

By law, U.S. workers are allowed to be exposed to 115 decibels for only 15 minutes a day. Anything over 140 can cause instant, irreversible hearing loss. The severity of the hearing loss is proportional to the length of exposure. A study from Ball State University in Indiana was typical in its finding that most long-time hunters who didn't use ear protection had worse hearing than the general population. The hearing loss isn't exactly peaceful, either. Gunfire-induced hearing loss is often accompanied by tinnitus, or ringing in the ears, and echoes. All this from shooting a few deer a year with a 140-decibel rifle. Now think of poor Rambo with his steady stream of gunfire at 150–190 decibels, 10 to 100,000 times louder than a single shot of a hunting rifle.

The U.S. Army never required hearing protection until the late 1960s. Talk to those veterans who saw active duty during World War II and the Korean War. There is a very good chance that many are partially or completely deaf in the ear closer to the gun. This type of damage is not merely from growing old. Hunters and marksmen never got the message about ear protection until the 1970s. Hunters, especially older ones, have been slow to warm to ear protection. They figure they're only taking a few shots during hunting season. This is somewhat true, but it only takes a few shots to damage the ears. Ear protection is most needed at indoor shooting ranges, where the rounds are many and the sounds echo. Fortunately, shooting-range operators are usually strict about requiring ear protection. The gun owners that everyone is worried about are called "junk shooters" in rural America and Canada. These folks shoot several rounds of ammo after work or on the weekends in their own backyard, knocking cans off stumps and such. Rarely do they wear ear protection. It's a macho thing. That is why movies like *Rambo* are troubling. More than just Hollywood bravado, they present a false reality about gun noise.

Is it really such an issue? With about 50 million gun owners in the United States, yes. There have been no studies on the rate of hearing loss across a broad range of gun owners. Studies have been limited to hunters, police officers, soldiers, and those who frequent gun ranges. These groups of individuals are easy to monitor. There exists only anecdotal evidence of a proliferation of near-deaf junk shooters from reports of their ear doctors, who say these gun users had no idea what was causing their hearing loss and chronic tinnitus. If gun users do not frequent the shooting range, there is little chance to educate them about the dangers of noise. The backyard junk shooters may be responsible gun owners; they simply do not know the kind of damage they are doing to their ears. Unlike Rambo, who would have been rendered deaf after just one movie, these backyard gun users will experience a gradual level of hearing loss year after year from steady exposure to 140 decibels.

Shooting ranges also provide eye protection from spent shells and gunpowder dust. Even fewer people know about the damage they can do to their eyes from gunfire.

Rambo did catch a small break with all that gunfire. He never fired a gun properly, raising it up to shoulder height to brace against the recoil. This kept the gun away from his ears and saved him a few decibels. Of course, firing the way he did is physically impossible. Rambo fired M16s—the type of automatic weapon that is usually bolted to the side of a ship or plane—with his one unbraced arm. Action, reaction: you may have seen folks having fun propelling themselves on an office chair by placing a fire extinguisher on their lap and firing it in the opposite direction. Well, the recoil from an M16 could rip Rambo's arm from his shoulder socket.

Even little handguns have a kick. Rarely in real life would you see a police officer shoot with one hand. (Actually, rarely in real life would you see a police officer shoot at all; most never use a gun on duty throughout their entire career.) Shooting without bracing can cause serious wrist strain or even snap your wrist bones, depending on the recoil strength and the angle of your wrist. A common injury among drug dealers trying to act cool, mimicking Hollywood, is breaking their thumbs while shooting semiautomatic weapons improperly, sideways with their thumb on the sliding cartridge. Shooting an automatic weapon with one arm would spin you around, no matter how strong you are, sending off a 360-degree hail of bullets. Depending on the weight of the gun, the recoil could easily dislocate your shoulder. In fact, soldiers occasionally do dislocate their shoulders from shooting assault rifles improperly.

Thus, shooting a handgun with any hope of accuracy with one arm out the car window while driving is nearly impossible. You couldn't hit the proverbial broad side of a barn. The recoil would send the gun up in the air. Rambo avoids the accuracy issue by spraying the entire forest with bullets. He easily went through several tens of thousands of bullets per movie. Where does this leave you? Ducking for cover, that's where. Another feature of Hollywood magic is that the bullets never ricochet. Hollywood bank robbers burst in to First National and fire into the ceiling. Just where do those bullets go? Plaster isn't flesh; it cannot absorb bullets so easily. What goes up must come down. One bullet toward the ceiling could easily ricochet down onto a marble countertop, then ricochet gut-level across the room. A bank robber takes a big risk when he fires and screams, "This is a stick-up!" He could nick himself or any of the bank's customers.

The jungle canopy is a treacherous place for gunfire. Depending on the grazing angle of the bullet, even delicate leaves and the surface of the water can send bullets off in different directions. Given the number of bullets that flew back and forth, Rambo probably nailed at least one of his own men. With the severe, noise-induced hearing loss, the rubbery and functionless arms, and mass mutiny from his own men over gross irresponsibility, Sylvester Stallone should be happy that Hollywood retired the Rambo character.

40

೬ ೬ ೬ ೬ ೬ ೬ ೬

Knocked Out, Loaded: Imagined Violence and Real Problems

My name is Bond, something Bond. I think. You'll have to forgive me. I've been rendered unconscious so many times that I suffer from long-term memory loss." That's certainly not the coolest of Bond lines, which is why you'll never hear it. That and "Sorry, honey, I meant to clean that vomit off my shoes." Getting knocked out is nasty business. Once isn't so bad. Yet each successive blow to the head multiplies your risk of suffering permanent memory loss, brain damage, or even a loss of vision, hearing, smell, or dexterity. A history of concussions may result in slurred speech, learning disabilities, balance problems, and emotional instability. No one can be certain whether boxer Muhammad Ali's Parkinson's disease stems from years of blows to the head. Boxers, however, suffer from neurological disorders at a far greater rate than the general population. You don't just shake off a blow to the head and move on. Head injuries linger.

A concussion is literally a bruise on the brain and a paralysis of the brain's nervous function caused by a blow to the head. The blow will nearly always cause confusion and short-term memory loss (a mild concussion) but not necessarily a loss of consciousness (the classic concussion). The risk of permanent brain damage grows greater with each concussion, regardless of the type, particularly when the damaging blows occur close to each other. Regardless of the timing, after the first blow, the chance of a second concussion after the next blow is four times greater. Even the mildest of concussions may cause flulike symptoms of dizziness and fatigue for a week. More serious concussions (whether knocked out or not) may lead to irritability and restless sleep for up to six months. This would certainly wreak havoc with 007's sex life.

High schools are finally taking concussions seriously after tragic reports of kids dying on the football field when they return to a game too soon after a blow to the head. The Centers for Disease Control and Prevention estimate that there are 200,000 concussions per year in high school sports. State by state, there are guidelines that characterize concussions into grades, the lowest grade being the least severe (with no amnesia or loss of consciousness) and the highest grade being the most severe (loss of consciousness). A grade 1 concussion forces a player out of the game for at least twenty minutes, pending further evaluation. A grade 2 concussion keeps a player out of the game and practice for at least a week; a higher grade concussion benches the player for at least a month and often for the year. Get knocked out twice, and you won't be allowed to play again. Football players are stereotyped as being dumb; there is no truth to this, but mild concussions aren't helping.

Tackling with a helmet on is one thing. What goes on during a Hollywood bar fight is far, far worse. Shattering a bottle or chair over someone's head is tantamount to murder. This isn't fun and games. It may seem obvious to you, but it is clearly not obvious to real-life drunken morons in bars who are motivated to commit violence because someone enters wearing the wrong professional sports jersey. First, bottles and chairs don't necessarily break so easily. The skull often breaks first. Second, when bottles and chairs do break, they are kind of sharp. So skulls, faces, arms, and hands tend to slice open. Third, the brain is kind of an important organ. Lots of blood flows up there to keep the brain nourished with oxygen and glucose. Breaking open the skull is a bloody mess.

The aftermath of a classic concussion is usually worse than that of a mild concussion. Confusion and disorientation will last longer.

Nausea accompanied by vomiting is just about guaranteed. Depending on where the blow hits, the victim could lose the sense of taste and smell. In fact, head injuries are a major cause of smell and taste disorders. Those who do not recover consciousness within a minute may lapse into a coma for days, weeks, or even months before regaining full consciousness. Remaining unconscious for over a day may lead to debilitating intellectual, emotional, or psychological problems. Blood clots caused by the blow may lead to a stroke if not identified and removed. No one wakes up from a bout of unconsciousness of any length without severe grogginess, a confusion that will leave a person unable to drive off, make getaway plans, or fight a villain.

Boxers clearly are the most susceptible to brain damage. Parkinson's disease is not known to be caused or triggered by blows to the head. The fact that boxing legend Muhammad Ali has this disease could be nothing more than a coincidence. The National Parkinson's Foundation estimates that less than 1 percent of the people who sustain significant head injuries will develop Parkinson's. Boxers are not so lucky in avoiding dementia pugilistica (punch-drunkenness) and chronic encephalopathy (chronic brain injury). Most boxers have at least some degree of brain damage, as revealed by MRI scans. Various studies show that anywhere from 15 to 40 percent of boxers have noticeable symptoms of brain damage: slurred speech, slow movements, confusion, random blackouts, emotional instability, severe mood swings, and confrontations with the law, to name but a few. Jack Dempsey and Joe Lewis were two famous boxers with visible symptoms of brain damage. Countless sparring partners and third-rate boxers with unscrupulous managers suffered a similar lot over the years. Even Mike Tyson displays the classic symptoms of brain damage, and he is usually the one delivering the knockout punch. Studies of boxers show a clear progression connecting brain damage with length of career and number of bouts. The brain disorders fall into four categories of diseases *similar* to multiple sclerosis, Alzheimer's disease, Parkinson's disease, and syphilis of the brain.

Nasty. One usually doesn't get this sense of danger from the movies. Fight scenes, no matter how serious, often come across as comical or at least entertaining. Even the weakling underdog can sneak up behind a hulking bully and render him unconscious with a beer bottle. As the dust settles, the bruised but unbloodied fighters slowly wake up, shake their heads (you can supply the sound effect), and continue with their day. No dizziness, no vomiting, no daylong confusion, no need for stitches and prompt medical emergency care, no unannounced bouts of vertigo for years to come. You'll find few exceptions to this in the world of cinema. *Thunderbolt and Lightfoot,* the 1974 film with Clint Eastwood and Jeff Bridges, is one exception. The seriousness of a head injury is actually woven into the story line. Lightfoot (Jeff Bridges) dies from an unstated neurological disorder a week or so after a serious blow to the head during a fight. An educated guess would be that one or more blood clots caused one or more mild strokes or placed excessive pressure on delicate regions around the brain. Bridges was nominated for an Academy Award for the role but lost to Robert DeNiro, a man who could throw and take a punch . . . Hollywood style.

41

⚕ ⚕ ⚕ ⚕ ⚕ ⚕

Heartbreaker: Hollywood Style

Bad acting meets bad medicine in the Hollywood heart attack. Folks at the American Heart Association and the National Heart, Lung and Blood Institute at NIH actually have educational campaigns aimed at erasing that silly image of a middle-aged fat guy grabbing his heart and keeling over. First, the thin and seemingly healthy nonsmoking athletic types can also have heart attacks. Second, only occasionally will a heart attack hit like a sharp pain in the heart. Rarer still will it cause instant death. Yet these reinforced stereotypes—fat guy, pain in the heart, keel over and die—act only to confuse the victim, the victim's family, or other bystanders in recognizing a heart attack and calling 911 for emergency assistance.

Of the million-plus heart attacks each year in the United States, over half are fatal because the victim doesn't seek medical treatment in time. Time is crucial. A variety of therapies can reduce the death rate by 25 percent if administered soon after symptoms start. The average delay in getting to a hospital is two hours, and a quarter of heart attack victims wait over five hours, according to a recent study published in the *American Heart Journal*. A measly 20 percent of the victims thumb a ride with an ambulance. These are the folks who understand they are having a heart attack. Around 10 percent of the heart attack victims drive to the hospital themselves while they are experiencing the symptoms, unaware they are having a heart attack, according to a survey published in *Circulation: Journal of the American Heart Association*. These are the folks who must own a Blockbuster video card.

The major signs of a heart attack, according to the American Heart Association, may be a mix of the following: uncomfortable pressure in the chest; unexplained sweating, nausea, shortness of breath, lightheadedness, or fainting; a pain spreading to the shoulders and arms or to the neck and jaw; a feeling of fullness; numbness in the arms; and that telltale pain in the center of the chest, lasting for several minutes. Other symptoms include paleness, weakness, and stomach or abdominal pain. Indeed, it is the stomach connection that makes some people think they are experiencing indigestion or heartburn and not a heart attack. Still, I don't know what pizza topping will cause shortness of breath and numbness in the arms.

Different folks will have different symptoms. Diabetics experiencing a heart attack may only encounter dizziness and sweating because the nerves around a diabetic's heart may not sense pain so well. People with angina or chronic chest pains may need to take nitroglycerin pills a couple of times to see if the pain disappears. Because angina patients often have chest pains, the key heart attack symptoms may be lightheadedness, sweating, nausea, or shortness of breath, in addition to chest pain. Women almost never experience a sharp pain in the chest during a heart attack. For women— who have longer delay times than men in getting to a hospital, and who often incorrectly believe they are at greater risk of breast cancer than cardiovascular disease—the symptoms can be everything *but* the sharp chest pain.

One almost accurate Hollywood heart attack was acted out by Marlon Brando in *The Godfather*. The largely improvised death scene finds the man disoriented, weak, and short of breath, before he finally collapses in his vegetable garden. In this case, there was a method to his method.

EPILOGUE

Tomorrow's Promise: Bad Medicine on the Horizon

T hings are getting better, right? Well, maybe not in the short run. At the turn of the millennium—when people thought Bill Gates's inability to create an operating system that recognized the difference between the years 1900 and 2000 would trigger Armageddon—the U.S. Centers for Disease Control and Prevention published a list of the top ten public health achievements in the twentieth century. At the dawn of the twenty-first century, most of these achievements are under attack.

The achievements, without ranking, are: vaccination, safer and healthier foods, the fluoridation and chlorination of drinking water, safer workplaces, the control of infectious diseases, motor-vehicle safety, the decline of cardiovascular diseases, family planning, healthier mothers and babies, and the recognition of tobacco as a health hazard. You may have already guessed some of the problems. Contentment breeds ignorance. Vaccinations are being undermined by a variety of camps—from the best educated and the "all naturals" to conspiracy mongers who think the government is out to poison them. These people simply don't understand what life was like before vaccines, when childhood diseases wiped out entire families.

Food safety has come a long way from the days a hundred years ago when most milk was tainted or watered down; most butter wasn't butter (I'll spare you the details); most meat was preserved with cancer-causing nitrates; canned food was prone to botulism;

poorer sections of the city had "second-hand" meat shops; and, in the big city, fresh vegetables were a rarity. Refrigeration and faster transportation solved most of these problems, along with food safety laws enacted during Teddy Roosevelt's presidency. Today we are cocky, relying on refrigeration and air travel to establish centrally located food-processing centers, leading to the rapid demise of local food production. Once again, a hundred years after Roosevelt's presidency, most meat is contaminated with harmful bacteria—deadly to those with weakened immune systems. Not a month goes by without some major recall of meat or vegetables. Megasized slaughterhouses and distribution centers, by virtue of the sheer bulk of food passing through them, are the breeding grounds for nationally distributed food-borne illnesses.

Fears that chlorine and fluorine cause cancer are prompting municipalities to lower or eliminate the amount of these elements in the drinking water supply. Fluoridation, once thought to be a Commie plot for mind control, has nearly eliminated tooth decay in the United States, a drop of close to 80 percent among children since its (slow) introduction in 1945. Fluorine's quality-of-life benefits, and its prevention of oral diseases that lead to oral cancers, dwarf its extremely low risk of bladder cancer. Likewise, without chlorinated water, bacteria would kill tens or even hundreds of thousands of people each year.

Workplaces in the United States are much safer now, although nonunionized blue-collar workers, whose numbers are rising with the dwindling popularity of unions, are at constant risk of workplace hazards. Walk past any small construction site with so-called day laborers—mostly undocumented immigrants or transients trying to make an honest buck—and you will find that few if any working the jackhammers are wearing ear protection. Look for ventilators, protective clothing, or even hard hats, and often you will not find them. Immigrant, nonunionized workers in U.S. meat processing plants face the constant threat of dismemberment, all for minimum wage. The response of U.S. businesses to stricter labor laws has been twofold: hire nonunionized workers too poor, desperate, or undereducated to know their rights; or move the company to a developing nation where worker safety laws don't exist. Also, if the United States continues the trend of burning more coal and nuclear fuel, this will lead to inevitable deaths and illnesses from coal and uranium mining, two of the riskiest jobs going.

Motor-vehicle safety is flying over the guardrails with the inane notion that bigger equals safer. Fifty percent of the cars sold in the United States are light trucks: pickups, vans, and SUVs. Big cars are death machines for the guy driving smaller, fuel-efficient vehicles. Big cars are harder to handle and make bigger impacts than do smaller cars. We seem to have entered into a transportation arms race, where safety is guaranteed as long as you have a bigger car than your enemy. Also, highways have become increasingly dangerous in the last few years as traffic congestion has doubled and even quadrupled in major cities. The demand for speed—depicted so skillfully in car commercials and remembered from the old days—and the confrontation of gridlock have led to epidemic levels of aggressive driving and road rage. Then add cell phones to the picture. Highway safety will get worse before it gets better.

The incidence of cardiovascular disease has declined over the past twenty years, but most experts fear this will shoot right back up in another ten years with the impact of overweight and obese baby boomers. Middle-aged folks today are not as healthy as their parents were twenty to thirty years ago. The impact on the faltering health care system will be huge. Everyone now knows that smoking is unhealthy, but 20 to 25 percent of us still smoke, and lung cancer is still the deadliest of all cancers. Tobacco-related illnesses will affect developing nations the most, as the tobacco industry turns to these countries for profit now that they have been duly vilified in the United States. Infectious diseases will start to climb a little as bacterial diseases such as tuberculosis become resistant to antibacterial medication, a result of overprescribing antibiotics. Transcontinental travel has brought diseases once confined to developing nations to the shores of America. Global warming will likely lead to increases in mosquito-borne diseases in North America, such as West Nile, dengue, and, in the southernmost parts, malaria, which was responsible for most of the deaths during the building of the Panama Canal.

The rates of infant and maternal mortality are down 90 percent and 99 percent, respectively, since 1900, a huge improvement. Still, the United States ranks at the bottom of this category among industrialized nations. Drug-addicted mothers continue to be the major cause of infant death and illness, although things are a little better since the crack-cocaine epidemic of the 1980s. Family planning is always in jeopardy. Access to contraceptive services has dramatically improved women's health and social standing worldwide. In 2001, the Bush administration reinstated the Reagan ban on international family planning funds; disallowed states to expand family planning services to the poor; proposed scrapping the plan for mandatory contraceptive coverage for federal employees; and pushed forward with an abstinence-only campaign.

Gingerly we step into a healthy future. The life expectancy rate is likely to fall by a few years in the United States with the rising number of obese children and a dependency on useless alternative therapies. There are no miracle cures on the horizon. Prolonged life will best be attained, in the near future, through a combination of preventive medicine (diet and exercise) and advances in bio-imaging and early disease detection, making the treatment of cancer and other diseases much more successful.

Genetics and stem cell research offer great promise, but we may be several decades away from miracle cures. Fear is our biggest obstacle. People confuse gene therapy with cloning, but the two have nothing to do with each other. (More on this issue below.) Stem cells, taken from a zygote (a fertilized egg, usually before it develops into an embryo), have the ability to develop into any number of human cells: nerve cells, blood cells, skin cells, and so on. The trick is coaxing the stem cells to turn into one specific type of cell. We are very close to understanding how to do this. The hope is that we could sprinkle a handful of stem cells along, say, the spinal cord of a paralyzed person and they would grow into new nerve cells and restore movement. Or, we could give sufferers of Parkinson's, Alzheimer's, ALS, or multiple sclerosis a new lease on life by supplying new cells that can multiply and replace dead or dying muscle and nerve cells. Many people, however, believe that the zygote is tantamount to human life and therefore cannot be sacrificed for experimentation and therapy. In the United States, the Bush administration agreed with this belief and essentially barred funding for stem cell research. Unless Europe and Asia can move forward with this promising and new field of research, the

As recently as 100 years ago, epidemics came swooping into an unsuspecting city to kill thousands of people in a matter of weeks. Folks who want to abandon vaccinations, chlorination, and other public health successes have no concept of how miserable life was in America . . . and still is for half the world. *Illustration by A. Paul Weber. Courtesy of the National Library of Medicine*

field will stall for four or more years. This is a huge blow to millions of people with horrible diseases—one spark of glimmering hope extinguished as quickly as it appeared.

Cloning is not what most people think it is. Clones are not beings that think, look, or act the same way as the original. Clones merely share the same DNA and are identical only at the moment of conception. After that, the clones set off on different paths—influenced by different chemicals and nutrients in the womb and shaped by different life experiences after birth. Identical twins are clones. Separate them at birth, and they may look radically different forty years later. Keep them together after birth, and they may think and act in radically different ways, despite their parents' desire to dress them alike. The premise of the movie *The Boys from Brazil*, although frightening and scientifically accurate to an interesting degree, cannot come true. We cannot clone Adolf Hitler. Correction: We may be able to clone him from whatever remains there are, but this human being will merely be Hitler's twin, who will have no greater chance of becoming an evil mastermind than you or I. Hitler's madness was not due to genes; rather, it was a unique product of chemicals, nutrients, acquaintances, complete strangers, tiny day-to-day events, and the gross socioeconomic environment of postwar Germany. This could never be reproduced.

Thus, those who demand that human cloning be legalized in the United States because their religion dictates that they be cloned to secure immortality are (a) crazy, (b) liars, or (c) both. Human cloning *should* be banned for now. We still do not know how to clone sheep and pigs that well. Many of these creatures die before reaching adulthood, and we don't fully understand how normal a life the animal survivors have. We should never subject a human baby to such chance. Consequently, just about every government in the world that has the technology to make a human clone has banned human cloning. The only ones in favor of cloning are the aforementioned idiots who claim this is part of their religion (an ancient religion born a few years ago). One can only hope they will not succeed by virtue of their being idiots, but that hasn't stopped others.

Gene therapy is not cloning. Rather, gene therapy works hand in hand with the Human Genome Project, the quest to map *and* understand the entire human genome—some fifty thousand genes spread out over forty-six chromosomes and wrapped up in a DNA molecule. These genes influence how we look, and how we respond to disease and medication. Genes bark out orders in the form of proteins; this is how they get cells to do things.

Through gene therapy, doctors hope to replace faulty genes with working models to cure genetic diseases. This endeavor is tantamount to transplanting microscopic molecules into trillions of cells—not an easy task. Dr. W. French Anderson, of the University of Southern California, conducted the first gene experiment in 1990. Since then, doctors haven't had much success with their microtransplanting. No one suffered from gene therapy, though, until 1999, when an eighteen-year-old volunteer with a manageable genetic disorder died in a gene therapy experiment. The experiment used a weakened cold virus, called the adenovirus, packed with "replacement" genes. The cold virus invades the body, as it always does, and is eventually killed by the body's own natural defense system. But before the cold virus dies, it transports the healthy gene into a target cell. Unfortunately, the cold virus turned worse and killed the volunteer. This tragic incident has chilled human gene therapy experiments for the moment.

Gene therapy cures for Alzheimer's and even heart disease, the top killer in the United States, are still several years away, but, as most researchers say, they are inevitable. The only thing that will knock progress off its track will be unfounded fears that gene therapy is tantamount to the creation of Frankenstein's monster. Nothing could be further from the truth. Doctors are merely replacing defective genes with good ones so that the body can heal itself of crippling diseases.

In the meantime, nothing will change for the youngest and the poorest in the United States and, perhaps, worldwide. The major killer among children will be gun violence and accidents. No one seems to be making any progress here. The poor will continue to have limited access to health care, so their symptoms of cancer, heart disease, infections, and diabetes will go undetected and lead to an early death. Over 90 percent of the time, early detection and treatment can cure disease—even many cancers. Cures exist to keep most people alive and healthy to an old age, yet we fail to use them, for social or economic reasons. In this respect, bad medicine may endure through the twenty-first century.

We tend to laugh at the way people approached health five hundred or a thousand years ago. I often wonder what the joke will be in the year 2500 when future societies look back at us. Surely chemotherapy will be considered the "bloodletting" of the twentieth and twenty-first century. Our approach to treating cancer—for lack of a better method—is to poison the cancer along with the entire body and hope the body can survive. This is proof positive that we don't know what the heck we're doing. Sure, we know about cells and DNA and proteins and chemical messengers. But we do not know how to regulate them. In many ways, we are as helpless as the cave dwellers were. Doctors in the future, we hope, will know how to efficiently isolate and remove cancerous cells or prevent them from growing in the first place. Immunology may be called into question, too, as it is refined to guarantee successful cures of bacterial and viral infections.

I believe that the twentieth century, in the mind of future historians, will blend into the fifteenth through nineteenth centuries as a time of enlightenment tinged with quackery. Lines will be drawn between prehistory, ancient China and Egypt, ancient Greece and Rome, a prolonged "dark era," and a five-hundred-year renaissance that started around 1500 and continues through to this day. Historians in centuries to come will say that humans from Descartes to Pauling, Watson, and Crick were on the right track—the way we view Hippocrates and Aristotle today. Historians will also smirk and amaze their students with tales of how a U.S. senator in the twentieth century could pass a law protecting homeopathy from legal scrutiny, or how the National Institutes of Health, which was then considered the most important medical establishment on earth, had a director who advocated homeopathy. Historians will pore through media clippings and remnants of twentieth-century pop culture to learn how wealthy Americans paid thousands of dollars to learn how they could improve their lives through ancient Indian and Asian practices that had long been dismissed in the very societies that developed them. Historians will learn of cycles of health crazes, such as homeopathy, magnet therapy, and gemstone healing from the seventeenth straight through the twentieth century and beyond. All in all, we will be classified as a superstitious people who attributed horrible scourges such as AIDS and even terrorism to God's wrath over homosexuals, as the influential Rev. Jerry Falwell does today; or who attributed diseases in general to imaginary forces, personality types, or astrological alignments, as purported by popular and influential teachers of Ayurveda, falun gong, and external qigong today.

The astute historian five hundred years from now will find little to differentiate the fifteenth century from the twentieth century in terms of medical advances, as broad as they seem today. We continue to live in an era of both good and bad medicine, as our ancestors did before us. The twenty-first century could mark the start of the new era. Are we are confident enough to embrace it?

APPENDIX

More Bad Medicine

No tour of bad medicine would be complete without a parting glance at this group of pesky medical misconceptions, for—with apologies to Irving Berlin—the malady lingers on.

Table 3—Too Bad Not to Mention

The Misconception	The Reality
You have a stomach flu.	There's no such thing as a stomach flu. The flu is caused by a virus attacking the respiratory system. Bacteria are likely bugging your tummy.
We only dream in black and white.	Close your eyes. If you can think in color, you can dream in color. Just ask Dorothy about her trip to Oz.
Dreams have deep meaning.	Maybe, but no one knows what dreams mean. Dreams are largely a way for your brain to file memories from the preceding day and mentally prepare for the next. A dream interpreter's stance that "horses mean strength; seagulls mean hope" is pure folly.
Greasy food causes acne.	Surprisingly, there is no connection between the amount of junk food you eat and acne. A serious nutritional deficiency can cause blemishes, but by this point you would also have rickets.

(continued)

Table 3 *(Continued)*

Aluminum poisoning causes Alzheimer's disease.	Nope. The theory has been tested to death. Those people most exposed to aluminum (metal workers, people who must take daily antacid medication that contains aluminum) are at no higher risk for Alzheimer's. Some Alzheimer's sufferers have aluminum deposits in their brain; most, however, do not. The cause of Alzheimer's is unknown.
Liposuction is healthy and safe.	Liposuction is purely a cosmetic procedure, more dangerous than most one-day surgical procedures. Recovery is very painful too. The removed fat is the harmless variety from under the skin, not the harmful fat that coats organs and sticks to arteries. Very little fat can be removed; this is not a way to lose weight.
An aspirin a day keeps the doctor away.	Aspirin has been proven to help prevent heart attacks and ischemic strokes in people who are at high risk for them. But aspirin can also have serious side effects. It is not a vitamin to pop every day for good heart health if you are healthy or even at moderate risk for heart disease. Check with your doctor about whether daily aspirin will do you more harm than good.
My kidneys are bursting.	When you feel the urge to urinate, it's your bladder that is expanding and sending signals to your brain. No urine accumulates in the kidneys.
Doctors are smart.	Maybe real doctors are. The world of alternative medicine is full of unqualified practitioners with titles that include the word *doctor*. Their degrees were either purchased from mail-order diploma mills or attained from nonaccredited institutions, often from outside the United States. Doctors that you should question include those that go by: Doctor of Naturopathy (N.D.), Naturopathic Medical Doctor (N.M.D.), Doctor of Natural Health (N.H.D.), Doctor of Eclectic Medicine (MDE), Fellow of the American College of Naturopathy (FACN), and Doctor of Philosophy in Natural Health or Doctor of Philosophy in Holistic Nutrition, both of which, unfortunately, are abbreviated as Ph.D.

(continued)

Table 3 *(Continued)*

Scientists are working on a cure for cancer.	Cancer isn't one disease; it is hundreds of different types of diseases caused by a variety of agents (bacteria, viruses, pollutants, ionizing radiation) that attack every part of the body differently. There will never be one cure. No one will ever win the Nobel Prize for curing "cancer."
Kidneys are being stolen from living people and sold on the black market.	Nope. This is pure urban legend. No one has ever awakened sore in a bathtub filled with ice after being kidnapped and having a kidney removed.
People are pricked with needles containing HIV, the virus that causes AIDS, attached to gasoline pump handles.	Nope. This is another urban legend. For one thing, viruses won't survive for more than a few minutes exposed to the elements.
Vivisection is cruel, and alternatives to animal testing exist.	There are, unfortunately, few alternatives to animal testing. Every miracle cure and procedure (penicillin, anesthesia, open-heart surgery) was first tested on animals. Would you undergo laser eye surgery or an organ transplant that wasn't tested? We can reduce the number of animals used in testing and use nonanimal methods to test cosmetics. But when it comes to finding cures to childhood diseases or other big problems, such as AIDS, we cannot do it without animals.

RECOMMENDED READING

꿈 꿈

The following list contains books, magazines, and web sites that will allow you to delve more deeply into the diverse health and anatomy topics presented in *Bad Medicine*.

Books and Periodicals

Stephen Jay Gould and Philip Kitcher have both taken aim at purveyors of pseudoscience. You really can't go wrong with any Gould book, and I recommend *The Mismeasure of Man* (W. W. Norton & Company, 1993). Kitcher's *Abusing Science* (MIT Press, 1986) systematically picks apart the antievolution movement. Likewise, S. Anthony Barnett's *Science: Myth or Magic* (Allen & Unwin, 2000) and Henry Bauer's *Science or Pseudoscience: Magnetic Healing, Psychic Phenomena, and other Heterodoxies* (University of Illinois, 2001) explore the reasons behind unfounded beliefs.

I thoroughly enjoyed Robert Park's *Voodoo Science* (Oxford University Press, 2000), in which Park takes readers down "the road from foolishness to fraud," getting into the minds of seemingly educated scientists hooked on homeopathy or perpetual motion machines. Sheldon Rampton and John Stauber of the Center for Media and Democracy have two humorously titled yet hard-hitting books about industrial practices: *Trust Us, We're Experts!* (Tarcher-Putman, 2001) and *Toxic Sludge Is Good For You!* (Tarcher-Putman, 1995). With these books, one can gain a sense of how science, statistics, and public relations are used and abused.

Leonard Hayflick wrote the definitive book on aging, *How and Why We Age* (Ballantine Books, 1996), which I recommend for its thoroughness, although it is tinged with fatalism. For the cheery side of aging, refer to Walter Bortz's *Dare To Be 100* (Simon & Schuster, 1996), and for antiaging scams, read S. Jay Olshansky's *The Quest for Immortality*

(W.W. Norton & Company, 2001). Navigating the world of alternative medicine can be treacherous. Verro Tyler's *The Honest Herbal* (Hawthorn Herbal Press, 1999), now in its fourth edition, is the bible of herbal medicine. I was surprised to find *Alternative Medicine for Dummies* (Wiley, 1998) to be rather responsible in explaining which therapies almost work and which are just silly, although the more specific alternative medicine books in the Dummies series (*Aromatherapy, Mind-Body Fitness*) frightened me.

Scientific American, Science News, the *New York Times* science section, and the *Washington Post* health section are all worth the subscription cost for those interested in keeping up with legitimate health and medical advances.

The World Wide Web

The World Wide Web is thick with bad medicine. Unfortunately, there are some very slick web sites presenting medical and health misinformation as if it were proven fact; you cannot necessarily judge a bad web site by its grammar mistakes and circa-1993 design. One valuable Internet resource, however, is Quackwatch (http://www.quackwatch.com), maintained by Dr. Stephen Barrett. Updated regularly, this site painstakingly combats health care fraud, myths, and fads. What the site lacks in hip design it makes up for many times over with its thoroughness. Another fine web site comes from the Committee for the Scientific Investigation of Claims of the Paranormal (http://www.csicop.org), which publishes *Skeptical Inquirer* magazine. Robert Todd Carroll maintains the Skeptic's Dictionary (http://skepdic.com), which provides an A-to-Z rundown of claims and rumors you perhaps have long wondered about.

For a more comical look at quackery, visit the Museum of Questionable Medical Devices (http://www.mtn.org/~quack), the Internet edition of a free, quirky museum in Minneapolis. Most will agree that the best inside joke among scientists is the Annals of Improbable Research, or AIR (http://www.improb.com), originators of the Ig Nobel Prize. AIR publishes results of bizarre and seemingly useless yet 100 percent real science experiments. Homeopathy is a regular feature. For the hardcore health fan, the National Institutes of Health (http://www.nih.gov) are a font of information if you dig deep enough. One useful site is the National Library of Medicine's PubMed service (http://www.nlm.nih.gov/hinfo.html), which offers a free abstract search of thousands of professional health journals.

BIBLIOGRAPHY

Introduction

BBC Online Education, "Medicine Through Time," http://www.bbc.co.uk/education/medicine.

Bettmann, O., *The Good Old Days—They Were Terrible!*, Random House, 1974.

Grope, R., "The Medicinal Leech," *Annals of Internal Medicine*, 1988; 108:399–404.

Hope, V. (ed.), *Death and Disease in the Ancient City*, Routledge, 2000.

Jouanna, J., *Hippocrates* (Medicine and Culture), Johns Hopkins University Press, 1999.

Laudan, R., "Birth of the Modern Diet," *Scientific American*, August 2000, 76–81.

Morens, D. M., "Death of a President," *New England Journal of Medicine*, 1999; 341:1845–1850.

Nuland, S., *The Mysteries Within: A Surgeon Reflects on Medical Myths*, Simon & Schuster, 2000.

Nunn, J., *Ancient Egyptian Medicine*, University of Oklahoma Press, 1996.

Shigelbrisa, K., "Interpreting the History of Bloodletting," *Journal of Historical Medicine and Allied Science*, 1995; 50:111–146.

Siraisi, N., *Medieval and Early Renaissance Medicine: An Introduction to Knowledge and Practice*, University of Chicago Press, 1990.

Part I. I Sing the Body Eclectic

Bower, B., "Brains in Dreamland," *Science News,* Aug. 11, 2001, 90–92.

"Brain Drain," Last Word (column), *New Scientist* 19, Dec. 26, 1998–Jan. 2, 1999.

Caldwell, S., and Popenoe, R., "Perceptions and Misperceptions of Skin Color," *Annals of Internal Medicine,* 1995; 122(8):614–617.

Cooper, R., and David, R., "The Biological Concept of Race and Its Application to Public Health and Epidemiology," *Journal of Health Politics, Policy and Law* 1986; 11(1):97–116.

Della Sala, S. (ed.), *Mind Myths: Exploring Popular Assumptions About the Mind and Brain,* John Wiley & Sons, 1999.

Diamond, J., *Guns, Germs, and Steel,* W. W. Norton & Company, 1999.

Goodman, S. M., "The Sin of Onan," *Journal of the Royal Society of Medicine,* 2000; 93(3):159.

Gould, J., *The Mismeasure of Man,* W. W. Norton & Company, 1993.

Kitcher, P., *Abusing Science: The Case Against Creationism,* MIT Press, 1982.

Makari, G. J., "Between Seduction and Libido: Sigmund Freud's Masturbation Hypotheses and the Realignment of his Etiologic Thinking, 1897–1905," *Bulletin of Historical Medicine,* 1998; 72(4):638–62.

Monell Chemical Sense Center, web site and personal correspondence.

Money, J., "The Genealogical Descent of Sexual Psychoneuroendocrinology from Sex and Health Theory: The Eighteenth to the Twentieth Centuries," *Psychoneuroendocrinology* 1983; 8(4):391–400.

Montagu, A., "Race: The History of an Idea," in *The Idea of Race,* University of Nebraska Press, 1965, 5–41.

National Eye Institute, "Common Myths and Old Wives' Tales About the Eyes," fact sheet, 2000.

National Institute on Aging, "Acne Fact Sheet," 2000.

National Institute on Deafness and Other Communication Disorders, Wise Ears! national educational campaign.

Norgen, R., "The Gustatory System," *The Human Nervous System* (Paxinos, E., ed.), Academy Press, 1990.

Powledge, T., "Head Games," *Washington Post,* April 10, 2001: H12.

Rusting, R., "Hair—Why It Grows, Why It Stops," *Scientific American,* June 2001: 71–79.

The Scan, "A Taste for Fat," *Washington Post,* Sept. 4, 2001: H03.

Smith, D., and Margolskee, R., "Making Sense of Taste," *Scientific American,* March 2001: 32–39.

Vander, A., et al., *Human Physiology,* McGraw Hill, 1990.

Venter, C., White House news conference on race and genes, July 28, 2000.

Vonnegut, K., *Galapagos,* Delacorte Press, 1985.

Wanjek, C., "National Initiative to Improve Minority Cancer Care," CBS HealthWatch by Medscape, April 6, 2000.

———, "Cancer Culture: Disease in Different Populations Studied," CBS HealthWatch by Medscape, June 15, 2000.

———, " 'Detoxifying' the Liver," *Washington Post,* Aug. 8, 2000: H08.

Whorton, J., "The Solitary Vice: The Superstition That Masturbation Could Cause Mental Illness," *Western Journal of Medicine,* 2001; 175(1):66–8.

Witzig, R., "The Medicalization of Race: Scientific Legitimization of a Flawed Social Construct," *Annals of Internal Medicine,* 1996; 125: 675–679.

Part II. Growing Old

American Association of Retired Persons, "Health and Wellness Guide," AARP online fact sheet, http://www.aarp.org.

Anderson, R., "U.S. Decennial Life Tables for 1989–91 Vol. 1, No. 4, 7–8 (National Center for Health Statistics, 1999)," *Nature,* Nov. 9, 2000, 267.

Anisimov, V., "Life Span Extension and Cancer Risk: Myths and Reality," *Experimental Gerontology,* July 2001; 36(7):1101–1136.

Begley, S., "Memory's Mind Games," *Newsweek,* July 16, 2001, 52–54.

Bortz, W., *Dare To Be 100,* Simon & Schuster, 1996.

Centers for Disease Control and Prevention, Chronic Disease Notes and Reports, *Healthy Aging,* CDC Newsletter, Fall 1999; (12)3:3–7.

Daniels, D., and Winter, W., "The Myth of Female Frailty, the Reality of Females and Physical Activity," *Recent Advances in Nursing,* 1989; 25:1–19.

Evans, G., "Sexuality in Old Age: Why It Must Not Be Ignored by Nurses," *Nursing Times,* 1999; 95(21):46–47.

Federation of American Societies for Experimental Biology, "Breakthroughs in Bioscience," FASEB Newsletter on osteoporosis, 2001.

Finch, C., and Austad, S., "History and Prospects: Symposium on Organisms with Slow Aging," *Experimental Gerontology,* April 2001; 36(4–6):593–597.

Fitti, J., and Kovar, M., "Supplement on Aging to the 1984 National Health Interview Survey," CDC Vital and Health Statistics, #21, Oct. 1987.

Gavrilov, L., and Gavrilov, N., *The Biology of Life Span,* Harwood Academic, 1991.

Gould, S., *Time's Arrow, Time's Cycle,* Harvard University Press, 1987.

Hayflick, L., *How and Why We Age,* Ballantine Books, 1996.

Jazwinski, S., "Metabolic Control and Ageing," *Trends in Genetics,* Nov. 2000; 16(11):506–511.

Kleinman, K., et al., "A New Surrogate Variable for Erectile Dysfunction Status in the Massachusetts Male Aging Study," *Journal of Clinical Epidemiology,* 2000; 53(1):71–78.

Kolata, G., "While Children Grow Fatter, Experts Search for Solutions," *New York Times,* Oct. 19, 2000, A1.

Kovar, M., et al., "The Longitudinal Study of Aging: 1984–90," CDC Vital and Health Statistics, #28, July 1992.

Leon, T., et al., "Effect of Dietary Restriction on Age-Related Increase of Liver Susceptibility to Peroxidation in Rats," *Lipids,* June 2001; 36(6):589–593.

McGlone, F., and Kick, E., "Health Habits in Relation to Aging," *Journal of the American Geriatric Society,* Nov. 1978; 26(11):481–488.

Miller, M., and Keller, T., "Measuring Drosophila (fruit fly) Activity during Microgravity Exposure," *Journal of Gravitational Physiology,* July 1999; 6(1):99–100.

National Institute on Aging, "Age Page" fact sheets.

Nature Insight, "Ageing," *Nature,* Nov. 2000; 408:231–269.

New England Centenarian Study, Harvard Medical School, online guide, http://www.bumc.bu.edu/Departments/HomeMain.asp?Department ID=361.

Padfield, A., "Myths in Medicine. Story That Early Retirement Is Associated with Longevity Is Often Quoted," *British Medical Journal,* June 1996; 312(7046):1611.

Perls, T., and Fretts, R., "The Evolution of Menopause and Human Life Span," *Annals of Human Biology,* May–June 2001; 28(3):237–245.

Perls, T., et al., *Living to 100,* Basic Books, 1999.

Rubin, J., and Tarrant, A. (producers), *Stealing Time,* PBS, June 2, 1999.

Scientific American Presents, The Quest to Beat Aging, Sept. 2000.

Shalala, D., Healthy People 2010 conference address, Washington, D.C., Jan. 25, 2000.

Specter, M., "Secret to Long Life in Azerbaijan? It's Not the Yogurt," *New York Times,* Mar. 14, 1998: A1.

Tyas, S., et al., "Risk Factors for Alzheimer's Disease: A Population-based, Longitudinal Study in Manitoba, Canada," *International Journal of Epidemiology,* 2001; (3):590–597.

Vaillant, G., and Mukamal, K., "Successful Aging," *American Journal of Psychiatry,* June 2001; 158(6):839–847.

Vander, A., et al., *Human Physiology,* McGraw Hill, 1990.

Wanjek, C., "Sight & Sound Loss Striking Earlier & Earlier," CBS HealthWatch by Medscape, March 15, 2000.

Wilcox, B., et al., *The Okinawa Program*, Clarkson Potter, 2001.

Wiley, D., and Bortz, W., "Sexuality and Aging—Usual and Successful," *Journals of Gerontology Series A: Biological Sciences and Medical Sciences,* 1996; 51(3):M142–146.

Wilmoth, J., "Demography of Longevity: Past, Present, and Future Trends," *Experimental Gerontology,* Dec. 2000; 35(9–10):1111–1129.

Woods, N., et al., "Memory Functioning among Midlife Women: Observations from the Seattle Midlife Women's Health Study," *Menopause,* 2000:(7)257–265.

Part III. Enough to Make You Sick

Barber, R., et al., "Oral Shark Cartilage Does Not Abolish Carcinogenesis but Delays Tumor Progression in a Murine Model," *Anticancer Research,* March–April 2001; 21(2A):1065–1069.

Brodeur, P., *The Zapping of America: Microwaves, Their Deadly Risk and the Cover-Up,* Norton, 1977.

————, *Currents of Death: Power Lines, Computer Terminals, and the Attempt to Cover Up Their Threat to Your Health,* Simon & Schuster, 1989.

————, *The Great Power-Line Cover-Up: How the Utilities and the Government Are Trying to Hide the Cancer Hazard Posed by Electromagnetic Fields,* Little, Brown & Co., 1993.

Carlo, G., and Schram, M., *Cell Phones: Invisible Hazards in the Wireless Age,* Carroll & Graf Publishers, Inc., 2001.

Dorahy, M., "Dissociative Identity Disorder and Memory Dysfunction: The Current State of Experimental Research and Its Future Directions," *Clinical Psychological Review,* July 2001; 21(5):771–795.

Ernst, E., and Cassileth, B., "How Useful Are Unconventional Cancer Treatments?" *European Journal of Cancer,* Oct. 1999; 35(11): 1608–1613.

Food and Drug Administration, "FDA Approved Celebrex in Adjunct Therapy for Familial Adenomatous Polyposis," FDA press release, December 23, 1999.

Gleaves, D., et al., "An Examination of the Diagnostic Validity of Dissociative Identity Disorder," *Clinical Psychological Review,* June 2001; 21(4):577–608.

Graunt, J., *Natural and Political Observations Mentioned in a Following Index, and Made Upon the Bills of Mortality,* Roycroft, London, 1662.

Hu, F., et al., "Diet, Lifestyle, and the Risk of Type 2 Diabetes Mellitus in Women," *New England Journal of Medicine,* 2001; 345:790–797.

Inskip, P., et al., "Cellular-Telephone Use and Brain Tumors," *New England Journal of Medicine,* 2001; 344:79–86.

Karlen, A., *Man and Microbes: Disease and Plagues in History and Modern Times,* Touchstone Books, 1996.

Keeling, M., and Gilligan, C., "Metapopulation Dynamics of Bubonic Plague," *Nature,* Oct. 2000; 407(6806):903–906.

———, "Bubonic Plague: A Metapopulation Model of a Zoonosis," Proceedings of the Royal Society of London—Series B: *Biological Sciences,* Nov. 2000; 267(1458):2219–2230.

Lalonde, J., et al., "Canadian and American Psychiatrists' Attitudes Toward Dissociative Disorders Diagnoses," *Canadian Journal of Psychiatry,* June 2001; 46(5):407–412.

Lane, W., and Comac, L., *Sharks Don't Get Cancer,* Avery Penguin Putnam, 1992.

———, *Sharks Still Don't Get Cancer,* Avery Penguin Putnam, 1996.

Mathews, J., "Media Feeds Frenzy over Shark Cartilage as Cancer Treatment," *Journal of the National Cancer Institute,* 1993; 85:1190–1191.

McDowell, I., "Alzheimer's Disease: Insights from Epidemiology," *Aging* (Milano), June 2001; 13(3):143–62.

Munoz, D., and Feldman, H., "Causes of Alzheimer's Disease," *Canadian Medical Association Journal,* 2000; 162(1):65–72.

Muscat, J., et al., "Handheld Cellular Telephone Use and Risk of Brain Cancer," *Journal of the American Medical Society,* 2000; 284:3001–3007.

National Cancer Institute, "Cartilage (Bovine and Shark)," NCI fact sheet, 2000.

Park, R., *Voodoo Science: The Road from Foolishness to Fraud,* Oxford University Press, 2000.

Pasteur, L., and Lister, J., *Germ Theory and Its Applications to Medicine & on the Antiseptic Principle of the Practice of Surgery,* Prometheus Books, 1996 (paperback reprint).

Pinn, G., "Herbal Medicine in Oncology," *Australian Family Physician,* June 2001; 30(6):575–80.

Titball, R., and Williamson, E., "Vaccination against Bubonic and Pneumonic Plague," *Vaccine,* July 20, 2001; 19(30):4175–4184.

Vander, A., et al., *Human Physiology,* McGraw Hill, 1990.

Wanjek, C., "Colon Cancer Exam Beckons at 50: Fear Not," CBS HealthWatch by Medscape, May 2000.

———, "Food Is Medicine in Colon Cancer Fight," CBS HealthWatch by Medscape, May 2000.

————, "Bare-Headed Lies," *Washington Post*, Jan. 16, 2001: H07.

————, "Cell Phones and Cancer, Nobody's Home," *Washington Post*, Feb. 6, 2001: H10.

Part IV. Eating It Up

Abelow, B., et al., "Cross-cultural Association between Dietary Animal Protein and Hip Fractures: A Hypothesis," *Calcified Tissue International*, 1992; 50:14–18.

Alpha-Tocopherol, Beta Carotene Cancer Prevention Study Group, "The Effect of Vitamin E and Beta Carotene on the Incidence of Lung Cancer and Other Cancers in Male Smokers," *New England Journal of Medicine*, 1994; 330:1029–1035.

American Heart Association, "Antioxidant Vitamins," online fact sheet, 2000, http://www.americanheart.org.

Atkins, R., *Dr. Atkins' New Diet Revolution*, Avon Health, 1992.

Barrett, S., "Antioxidants and Other Phytochemicals: Current Scientific Perspective," *Quackwatch*, Jan. 13, 2000.

Bratman, S., *Health Food Junkies: Overcoming the Obsession with Healthy Eating*, Broadway Books, 2001.

Brody, J., "Debate Over Milk: Time to Look at Facts," *New York Times*, Sept. 26, 2000: D8.

————, "One-Two Punch for Losing Pounds; Exercise and Careful Diet," *New York Times*, Oct. 17, 2000: D6.

————, "Fat but Fit: A Myth About Obesity Is Slowly Being Debunked," *New York Times*, Oct. 24, 2000: D7.

Cuatrecasas, P., et al., "Lactase Deficiency in the Adult: A Common Occurrence," *The Lancet*, 1965; 1:14–18.

Dixon, B., "The Bottled Water Boom," *British Medical Journal* (Clinical Research Edition), Jan. 1988; 296(6617):298.

Federation of American Societies for Experimental Biology, "Breakthroughs in Bioscience," FASEB Newsletter on osteoporosis, 2001.

Ferrier, C., "Bottled Water: Understanding a Social Phenomenon," for the World Wildlife Fund, April 2001.

Feskanich, D., et al., "Milk, Dietary Calcium, and Bone Fractures in Women: A 12-year Prospective Study," *American Journal of Public Health*, 1997; 87(6):992–997.

Flam, F., "Shopping for an Answer to Cancer," *Philadelphia Inquirer*, Nov. 13, 2000: C01.

Griffin, S., et al., "Quantifying the Diffused Benefit from Water Fluoridation in the United States," *Community Dental Oral Epidemiology*, April 2001; 29(2):120–9.

Halliwell, B., "Viewpoint: The Antioxidant Paradox," *The Lancet,* 2000; 355:1179–1180.

Hennekens, C., et al., "Antioxidant Vitamins: Benefits Not Yet Proved," *New England Journal of Medicine,* 1994; 330:1080–1081.

Hudson, V. (ed.), "Herbals: Therapeutic and Adverse Effects—A Bibliography with Abstracts," *National Library of Medicine,* July 2000.

Institute of Medicine of the National Academies, "Antioxidants' Role in Chronic Disease Prevention Still Uncertain; Huge Doses Considered Risky," press release, April 10, 2000.

Kolata, G., "Chemicals in the Brain Tell the Body 'It's Time to Eat,'" *New York Times,* Oct. 17, 2000: D1.

———, "No Days Off Are Allowed, Experts on Weight Argue," *New York Times,* Oct. 18, 2000: A1.

———, "While Children Grow Fatter, Experts Search for Solutions," *New York Times,* Oct. 19, 2000: A1.

Kushi, A., *Aveline Kushi's Complete Guide to Macrobiotic Cooking for Health, Harmony and Peace,* Warner Books, 1985.

Lalumandier, J., and Ayers, L., "Fluoride and Bacterial Content of Bottled Water vs Tap Water," *Archives of Family Medicine,* March 2000; 9(3):246–250.

Lanou, M., Physicians Committee for Responsible Medicine, personal correspondence and interview, Aug.–Sept. 2001.

Lindner, L., "Stone Age Soup," *Washington Post,* Feb. 13, 2001: H09.

———, "Whole Grains, Half Truths," *Washington Post,* July 31, 2001: H01.

MacIntosh, D., et al., "Dietary Exposures to Selected Metals and Pesticides," *Environmental Health Perspective,* Feb. 1996; 104(2): 202–209.

———, "Longitudinal Investigation of Dietary Exposure to Selected Pesticides," *Environmental Health Perspective,* Feb. 2001; 109(2): 145–150.

Menkes, M., et al., "Serum Beta-Carotene, Vitamins A and E, Selenium, and the Risk of Lung Cancer," *New England Journal of Medicine,* Nov. 1986; 315(20):1250–1254.

Millory, S., *Junk Science Judo,* Cato Institute, 2001.

National Heart, Lung and Blood Institute, "Summary Report—Working Group on Atheroprotective Genes," March 29, 2000.

National Institutes of Health, "Facts about Dietary Supplements—Selenium," NIH fact sheet, Office of Dietary Supplements, March 2001.

National Resources Defense Council, "Bottled Water: Pure Drink or Pure Hype?" March 1999.

Pip, E., "Survey of Bottled Drinking Water Available in Manitoba, Canada," *Environmental Health Perspective*, Sept. 2000; 108(9):863–866.

Pollan, M., "Naturally—How Organic Became a Marketing Niche and a Multibillion-Dollar Industry," *New York Times Magazine*, May 13, 2001, 30–37, 57.

Rapola, J., et al., "Randomised Trial of Alpha-tocopherol and Beta-carotene Supplements on Incidence of Major Coronary Events in Men with Previous Myocardial Infarction," *The Lancet*, 1997; 349: 1715–1720.

Ryan, P. B., et al., "Analysis of Dietary Intake of Selected Metals in the NHEXAS-Maryland Investigation," *Environmental Health Perspective*, Feb. 2001; 109(2):121–128.

Scrimshaw, N., and Murray, E., "The Acceptability of Milk and Milk Products in Populations with a High Prevalence of Lactose Intolerance," *American Journal of Clinical Nutrition*, 1988; 48:1083–1085.

Seelye, K., "Arsenic Standards for Water Is Too Lax, Study Concludes," *New York Times*, Sept. 11, 2001: 18.

Squires, S., "Gulp! Vitamin Facts," *Washington Post*, Nov. 28, 2000: H13.

————, "Hearts & Minds," *Washington Post*, July 24, 2001: H10.

Stauber, J., and Rampton, S., *Toxic Sludge Is Good for You!: Lies, Damn Lies and the Public Relations Industry*, Tarcher Putnam, 1995.

Tribble, D., "Antioxidant Consumption and Risk of Coronary Heart Disease: Emphasis on Vitamin C, Vitamin E, and β-Carotene," *Circulation*, 1999; 99:591–595.

Tsubono, Y., "Green Tea and the Risk of Gastric Cancer in Japan," *New England Journal of Medicine*, March 2001; 344:632–636.

United States Department of Agriculture, Agricultural Fact Book 1998.

————, "National Organic Program," 2000, Docket Number: TMD-00-02-FR; RIN: 0581-AA40.

United States Environmental Protection Agency, "The Role of Use-Related Information in Pesticide Risk Assessment and Risk Management," EPA Office of Pesticide Programs, Aug. 21, 2000.

Vander, A., et al., *Human Physiology*, McGraw Hill, 1990.

Wanjek, C., "Mixed Messages," *Washington Post*, Aug. 7, 2001: F1.

Weaver, C., and Plawecki, K., "Dietary Calcium: Adequacy of a Vegetarian Diet," *American Journal of Clinical Nutrition*, 1994; 59(suppl): 1238S-41S.

Part V. The Return of the Witch Doctor

American Heart Association, "Aspirin in Heart Attack and Stroke Prevention," AHA fact sheet, 2000.

American Society of Plastic Surgeons, Liposuction fact sheet, 1998.

Ang-Lee, M., "Herbal Medicines and Perioperative Care," *Journal of the American Medical Association,* July 2001; 286(2):208–216.

Barrett, S., "Homeopathy, the Ultimate Fake," *Quackwatch,* rev. June 8, 2001.

————, "Oxygenation Therapy: Unproven Treatments for Cancer and AIDS," *Quackwatch,* rev. June 17, 2001.

Bruner, J., and de Jong, R., "Lipoplasty Claims Experience of U.S. Insurance Companies," *Plastic and Reconstructive Surgery,* April 2001; 107(5):1285–1292.

Chopra, D., *Return of the Rishi: A Doctor's Story of Spiritual Transformation and Ayurvedic Healing,* Houghton Mifflin, 1991.

————, *Ageless Body, Timeless Mind: The Quantum Alternative to Growing Old,* Crown Publishing, 1993.

————, *Boundless Energy,* Harmony Books, 1995.

Coleman, W., et al., "Guidelines of Care for Liposuction," *Journal of the American Academy of Dermatology,* Sept. 2001; 45(3):438–447.

Coulter, H., *Vaccination, Social Violence, and Criminality: Medical Assault on the American Brain,* North Atlantic Books, 1990.

Cucherat, M., et al., "Evidence of Clinical Efficacy of Homeopathy: A Meta-analysis of Clinical Trials," *European Journal of Clinical Pharmacology,* April 2000; 5(1):27–33.

Curtis, S., *Essential Oils,* Aurum, 1996.

de Jong, R., "Body Mass Index: Risk Predictor for Cosmetic Day Surgery," *Plastic and Reconstructive Surgery,* Aug. 2001; 108(2):556–563.

Ernst, E., "Mistletoe for Cancer?" *European Journal of Cancer,* 2001; 37:(1)9–11.

Hall, D., *You Can't Catch a Cold,* iUniverse, Inc., 2001.

He, J., et al., "Aspirin and Risk of Hemorrhagic Stroke: A Meta-analysis of Randomized Controlled Trials," *Journal of the American Medical Association,* 1998; 280:1930–1935.

Hennekens, C., et al., "Aspirin as a Therapeutic Agent in Cardiovascular Disease," a Statement for Healthcare Professionals from the American Heart Association, 1997.

Herbert, P., and Hennekens, C., "An Overview of the 4 Randomized Trials of Aspirin Therapy in the Primary Prevention of Vascular Disease," *Archives of Internal Medicine,* 2000; 160:3123–3127.

Jacobs, J., et al., "Treatment of Childhood Diarrhea with Homeopathic Medicine: A Randomized Clinical Trial in Nicaragua," *Pediatrics*, 1994; 93:719–725.

Jonas, W., and Jacobs, J., *Healing with Homeopathy: The Complete Guide*, Warner Books, 1996.

Karmo, F., et al., "Blood Loss in Major Liposuction Procedures: A Comparison Study Using Suction-Assisted versus Ultrasonically Assisted Lipoplasty," *Plastic and Reconstructive Surgery*, July 2001; 108(1): 241–249.

Kent, C., and Gentempo, P., "Immunizations: Fact, Myth, and Speculation," *International Review of Chiropractic*, Nov.–Dec. 1990.

Linde, K., et al., "Are the Clinical Effects of Homeopathy Placebo Effects? A Meta-analysis of Placebo-Controlled Trials," *The Lancet*, 1997; 350 (9081):834–843.

Linde, K., et al., "The Methodological Quality of Randomized Controlled Trials of Homeopathy, Herbal Medicines and Acupuncture," *International Journal of Epidemiology*, June 2001; 30(3):526–31.

Miller, N., *Vaccines: Are They Really Safe and Effective?* New Atlantean Press, 1995.

Park, R., *Voodoo Science: The Road from Foolishness to Fraud*, Oxford University Press, 2000.

Reuters News Service, "FDA Warns Firms on Adding Herbs to Food, Drink," June 7, 2001.

Rosa, L., et al., "A Closer Look at Therapeutic Touch," *Journal of the American Medical Society*, 1998; 279:1005–1010.

Rose, P., *Magnetic Therapy Illustrated*, Ulysses Press, 2001.

———, *The Practical Guide to Magnetic Therapy*, Sterling Publishing, 2001.

Sampson, W., "Analysis of Homeopathic Treatment of Childhood Diarrhea," *Pediatrics*, 1995; 96:961–964.

Skolnick, A., "The Maharishi Caper: Or How to Hoodwink Top Medical Journals," *ScienceWriters: The Newsletter of the National Association of Science Writers*, Fall 1991.

Steuer-Vogt, M., et al., "The Effect of an Adjuvant Mistletoe Treatment Programme in Resected Head and Neck Cancer Patients: A Randomised Controlled Clinical Trial," *European Journal of Cancer*, 2001; 37(1):23–31.

Thrift, A., et al., "Risk of Primary Intracerebral Hemorrhage Associated with Aspirin and Non-steroidal Anti-inflammatory Drugs: Case-control Study," *British Medical Journal*, March 1999; 318:759–764.

Tyler, V., *The Honest Herbal,* 4th Ed., Hawthorn Herbal Press, 1999.

Vandenbroucke, J., "Homoeopathy Trials: Going Nowhere," *The Lancet,* 1997; 351(9099):365.

Wanjek, C., "News Flash: Herbal Supplement for Menopause Hits the Big Time," *Washington Post,* April 10, 2001: H07.

Warrier, G., and Gunawant, D., *The Complete Illustrated Guide to Ayurveda: The Ancient Indian Healing Tradition,* Element, 1997.

Wise, J., "Health Authority Stops Buying Homoeopathy," *British Medical Journal,* 1997; 314:1574.

Worwood, S., *Essential Aromatherapy,* New World Library, 1995.

Part VI. Risking It All

Barnwell, Y., "More Than a Paycheck," Barnwell's Notes Co., 1982.

Barrett, S., "Bernadean University: A Mail-Order Diploma Mill," *Quackwatch,* rev. Mar 2001.

Brodeur, P., *Outrageous Misconduct: The Asbestos Industry on Trial,* Pantheon Books, 1985.

Children's Defense Fund, "The State of America's Children Yearbook, 2001."

Coile, D., and Miller, N., "How Radical Animal Activists Try to Mislead Humane People," *Laboratory Primate Newsletter,* July 1984; 23(3):11–13.

Cragin, D., and Lewis, J., "Eating Candy for Longevity and Other Toxic Sciences," address to the National Capital Area Skeptics, Jan. 20, 2001. (My inspiration for this topic.)

de Forest, L., *American Chamber of Horrors: The Truth about Food and Drugs,* Farrar and Rinehart, 1926.

Dodds, W., and Orlan, F. (eds), *Scientific Perspectives on Animal Welfare,* Academic Press, 1982.

Hill, A., "The Environment and Disease: Association or Causation?" *Proceedings of the Royal Society of Medicine,* 1965; 9:295–300.

Itoh, N., et al., "Have Sperm Counts Deteriorated over the Past 20 Years in Healthy, Young Japanese Men? Results from the Sapporo Area," *Journal of Andrology,* Jan. 2001; 22(1):40–44.

Kamrin, M., *Toxicology: A Primer,* Lewis Publishers, 1988.

Laudan, L., *Danger Ahead: The Risks You Really Face on Life's Highway,* John Wiley & Sons, 1997.

Lee, I., and Paffenbarger, R., "Life Is Sweet: Candy Consumption and Longevity," *British Medical Journal,* Dec. 1998; 317(7174):1683–1684.

Lu, F., *Basic Toxicology,* Hemisphere Publishing Corp., 1991.

McCally, A., et al., "Corneal Ulceration Following Use of Lash-Lure," *Journal of the American Medical Association,* 1933; 101(20):1561.

Miller, N., "Values and Ethics of Research on Animals," *Laboratory Primate Newsletter,* July 1984; 23(3):1–10.

Millory, S., *Junk Science Judo,* Cato Institute, 2001.

Rampton, S., and Stauber, J., *Trust Us, We're Experts!,* Tarcher-Putman, 2001.

Redelmeier, D., and Singh, S., "Survival in Academy Award-Winning Actors and Actresses," *Annals of Internal Medicine,* May 2001; 134(10):955–962.

Rowan, A., *Of Mice, Models and Men,* State University of New York Press, 1984.

Stauber, J., and Rampton, S., *Toxic Sludge Is Good For You! Lies, Damn Lies and the Public Relations Industry,* Tarcher-Putman, 1995.

U.S. Environmental Protection Agency, "Health Assessment Document for Chloroform," EPA-600/8-84-004F, Aug. 1985.

———, "Questions and Answers about Dioxins," EPA fact sheet, July 2000.

Vander, A., et al., *Human Physiology,* McGraw Hill, 1990.

Wanjek, C., "United States Confronts Mexican Border Health Problems," CBS HealthWatch by Medscape, March 2000.

———, "US Struggles to Meet Asian-American Healthcare Needs," CBS HealthWatch by Medscape, March 2000.

———, "National Initiative to Improve Minority Cancer Care," CBS HealthWatch by Medscape, April 6, 2000.

———, "Cancer Culture: Disease in Different Populations Studied," CBS HealthWatch by Medscape, June 2000.

———, "The Unfriendly Skies of Medical Research," *Washington Post,* Nov. 28, 2000: H06.

Zurlo, J., et al., "Animal and Alternatives in Testing: History, Science, and Ethics," Johns Hopkins Center for Alternatives to Animal Testing.

Part VII. Just Like in the Movies

Casson, I., "Boxing and Parkinson Disease," for the National Parkinson Foundation, 2001.

———, "Brain Damage in Modern Boxers," *Journal of the American Medical Association,* 1984; 251:2663–2667.

Finnegan, J., "Mass Media, Secular Trends, and the Future of Cardiovascular Disease Health Prevention: An Interpretive Analysis," *Preventive Medicine,* Dec. 1999; 29:550–558.

Goff, D., et al., "Prehospital Delay in Patients Hospitalized with Heart Attack Symptoms in the United States: The REACT Trial," *American Heart Journal,* Dec. 1999; 138:1003–1004.

Hanke, C., et al., "Blast Tattoos Resulting from Black Powder Firearms," *Journal of the American Academy of Dermatology,* Jan. 1989; 20(1):137–138.

National Institute on Deafness and Other Communication Disorders, "Noise and Hearing Loss," NIH Consensus Statement, 1990; 8(1): 1–24.

National Institute on Deafness and Other Communication Disorders, Wise Ears! national educational campaign.

Nondahl, D., et al., "Recreational Firearm Use and Hearing Loss," Archives of Family Medicine, 2000; 9:352–357.

Ransford, M., "Hunters, Recreational Shooters Should Always Protect Hearing," Ball State University News Center, Oct. 14, 1997.

Vander, A., et al., *Human Physiology,* McGraw Hill, 1990.

Wanjek, C., "Concussion Is Not Part of the Game," CBS HealthWatch by Medscape, Nov. 1999.

Zapka, J., et al., "Missed Opportunities to Impact Fast Response to AMI Symptoms," Patient Education Counsel, April 2000; 40(1):67–82.

Epilogue. Tomorrow's Promises:
Bad Medicine on the Horizon

Centers for Disease Control and Prevention, "Ten Great Public Health Achievements—United States, 1900–1999," Morbidity and Mortality Weekly Report, April 1999; 48(12):241–243.

Wanjek, C., "Gene Therapy at the Crossroads," CBS HealthWatch by Medscape, Jan. 2000.

INDEX

᪥ ᪥ ᪥ ᪥ ᪥ ᪥